SEA LEVEL

OCEANS IN DEPTH

A Series Edited by Katharine Anderson
and Helen M. Rozwadowski

SEA LEVEL

A HISTORY

WILKO GRAF VON HARDENBERG

THE UNIVERSITY OF CHICAGO PRESS
CHICAGO AND LONDON

The University of Chicago Press, Chicago 60637

The University of Chicago Press, Ltd., London

Published 2024
Printed in the United States of America

33 32 31 30 29 28 27 26 25 24   2 3 4 5

ISBN-13: 978-0-226-83183-1 (cloth)
ISBN-13: 978-0-226-83459-7 (e-book)
DOI: https://doi.org/10.7208/chicago/9780226834597.001.0001

Library of Congress Cataloging-in-Publication Data

Names: Hardenberg, Wilko Graf von, author.
Title: Sea level : a history / Wilko Graf von Hardenberg.
Other titles: Oceans in depth.
Description: Chicago ; London : The University of Chicago, 2024. | Series: Oceans in depth | Includes bibliographical references and index.
Identifiers: LCCN 2024000349 | ISBN 9780226831831 (cloth) | ISBN 9780226834597 (ebook)
Subjects: LCSH: Sea level—History. | Sea level—Measurement—History.
Classification: LCC GC89 .H37 2024 | DDC 551.45/809—dc23/eng/20240131
LC record available at https://lccn.loc.gov/2024000349

♾ This paper meets the requirements of ANSI/NISO Z39.48-1992 (Permanence of Paper).

FÜR PAPI

CONTENTS

FOREWORD

OCEANS IN DEPTH

Oceans and their restless waters have long given us metaphors for both permanence and flux. "Sea level" is a powerful abstraction, fundamental to our view of nature—a baseline for every description of the earth's vertical features. Yet today, headlines emphasize an alarming change: an inexorably rising sea. Along with climbing temperatures and hurricanes, sea level is, in the twenty-first century, a key marker of environmental crisis.

The story of sea-level measurement illustrates how ocean history can be anchored in particular places but always moves to a planetary perspective. This is among the compelling puzzles *Sea Level: A History* invites us to consider: When we see coastlines changing, is it a local or a general phenomenon? What shapes the long, shifting relationship between land and sea? How can we distinguish between short-term fluctuation, slow response to geological processes, and rapid anthropocenic change? And because this a human history, involving politics and technology: Who has measured the sea, and to what ends?

Addressing these questions, Wilko Graf von Hardenberg lays out a surprising history of ideas but also a complex account of measurement itself over the last three hundred years. We learn how geological conceptions have intersected with the history of technical devices and systems, from local harbor records to global satellites. The challenges of measuring the sea reveal, as well, how institutions and communities build scientific consensus and respond to changing data. In the eighteenth century, the dominant idea was that sea level was dropping. By the early nineteenth century, geologists began to adopt a new theory of a stable sea. In the next decades, measuring sea level became increasingly essential to the engineers and administrators who built harbors, canals, and bridges, at home and in far-flung empires.

By the mid-twentieth century, the idea of a stable sea had collapsed: rising sea level became an increasingly robust scientific concept, while visions of a drowning world entered popular fiction. Today, our measurements of sea level continue to reveal a profound truth: understanding the natural world involves thinking at multiple scales, large and small. As Hardenberg explains, measuring sea level requires looking at the tides and their movements, scanning blips in satellite images, scrutinizing a global aggregate of data. Entangled with these pictures we find politics, infrastructure, and imagination. We can never focus on just one measurement or one perspective: sea level means all of these at the same time.

This history of sea level—as an idea and a practice—makes clear why historical ideas about oceans in three dimensions and on a planetary scale matter to the twenty-first century.

Katharine Anderson
Helen M. Rozwadowski

FROM HEIGHTS TO MUDS

The peak of Mount Everest, 8,849 meters above sea level. Chimborazo, 6,263 m; Mont Blanc, 4,806 m; Pradidali hut, at the foot of the Pala group in the Dolomites, 2,278 m. It is common practice to provide the elevation of a place as part of its coordinates. Simple readings are now readily available, as altimeters of varying precision are embedded in our phones, car navigators, and multitudes of wearables. But elevation is necessarily relative: only the choice of a reference point allows us to express numerically the altitude of an object or location. Change the frame of reference or the accuracy of an instrument, and even the apparently stable measure of a mountain is revealed as but a snapshot of a specific technological moment.[1]

Altitude measurements routinely use mean sea level as a baseline; in most countries the official height reference framework refers to it in some way. The idea of sea level as a benchmark for elevations has by now been around for so long as to go essentially unnoticed—we mention it without pausing to consider what it means. That the concept has a history is easily ignored. We tend to forget that sea level is—far from a natural index—a product of technically and culturally determined assumptions. In this book I tell a story of these assumptions.

Appropriately, the book was conceived during a balmy summer spent between sea and mountains. A month after visiting the French Riviera and enjoying the sandy beaches of Nice and Menton with the woman who is now my wife, I went trekking with friends in the Pala group. Resting on a pass near the Pradidali hut, we began discussing the readings offered by our altimeters. Why did they vary ever so slightly from one instrument to the next? What did the numbers actually refer to? Different maps likewise recorded different elevations—why? This sense of uncertainty in establishing altitude, short-circuited in my mind with both the recent memory of gently rolling Mediterranean tides and menacing headlines about rising sea levels, leading me to think about the meaning of zero. This is how I embarked on a quest for the conceptual history of sea level.

Human influence on sea levels is a fairly recent phenomenon. The awareness of it, even more so. But the means we use to assess the effects of anthropogenic climate change are, like those we use to determine elevation, rooted in history—in cultural, social, and scientific agendas that were radically different from current ones. Establishing the level of the sea is part of a broader striving to reform and unify reference points and units of measure that has taken place since early modernity. The definition of the meter, the choice of a prime meridian, and the standardization of time are exemplars of this process. The use of sea level both as a baseline for measuring change and as a reference point for altitude is intertwined with a long-held perception of holocenic stability. And, paradoxically, with its recent upheaval.

In this book I examine three crucial stages in the history of mean sea level: how, when, and why it was first defined; how it became the prime geodetic reference point for elevations; and, finally, how it has been, more recently, redefined as a powerful symbol of the Anthropocene, the proposed current geological age in which humans attain a dominant role in effecting climatic and environmental changes. These stages mirror the development of human conceptions of the sea, from epitome of natural stasis to barometer of ongoing change. Since I began thinking about this project in August 2011, the issue of sea-level rise and the need to place it historically have only become more urgent: between then and the end of 2022 the global sea rose by almost 5 centimeters, about half of the overall rise recorded by satellites since 1993.[2]

METHODS OF MEASURE

Extreme accuracy in determining heights was long seen as unnecessary. The first altitude recorded on a map appeared only in 1712, when the physician and mathematician Johann Jakob Scheuchzer indicated an approximate value for the height of the Steilerhorn in the Lepontine Alps of the Grisons on his famous map of Switzerland, the *Nova Helvetiae tabula geographica*. Scheuchzer assessed the elevation barometrically with respect to an unspecified location on the Mediterranean Sea. The conversion of barometer readings into altitude values was a recent scientific accomplishment and still rather imprecise: Scheuchzer gives the Steilerhorn's height as about 3,500 meters (12,000 *pedes* or Roman feet), while according to modern measurements its summit reaches just 2,980 meters.[3]

The earliest known barometers had been built just a few decades earlier as collaterals of scholars' attempts to prove the possibility of producing a vacuum. When the open end of a glass tube full of mercury is plunged in a bowl containing the same substance, a vacuum forms near the closed end. The effect of external air pressure keeps the level of mercury in the tube and in the bowl in balance. If air pressure rises, the mercury will rise in the tube. The first such tube to which a scale was attached—thus the first instrument truly to measure air pressure—was built by the Italian physicist Evangelista Torricelli in 1643. The device's potential as an altimeter was recognized almost immediately. Five years later the philosopher and mathematician Blaise Pascal, an expert in fluid mechanics and atmospheric pressure, asked his brother-in-law Florin Périer to bring a tube built according to Torricelli's design to the top of the Puy de Dôme in the French Massif Central. While the instrument was graduated and care had been taken to calibrate it, the experiment was only able to prove that air has weight, not, Périer lamented, to provide a rule for precisely assessing altitude. The formulas necessary to transform pressure values into heights with some precision took longer to develop. Crucial in this regard was the parallel improvement of methods and infrastructures for trigonometric surveys, which yielded reliable height assessments with which to compare barometric readings.[4]

While the principles of determining heights geometrically were known in both early China and the Mediterranean world, elevations had not previously been shown on maps. Recording the difference in height between peak and valley bottom, or simply the fact that one place lay higher than another, appeared sufficient for most purposes. The anonymous Chinese maker of the so-called Mawangdui maps, in the second century CE, represented the heights of the nine peaks of the

Jiuyi Mountains in Hunan province as simple bars. These showed the peaks' relative elevation and, possibly, climatic information. In Europe before the eighteenth century, the otherness and danger of mountain regions were foregrounded over the physical height of their peaks. With a handful of exceptions, such as the ascent, in 1336, of Mont Ventoux near Avignon by the Italian poet Francesco Petrarca, mountain summits served mostly as background. Greater focus was directed to passes or caves as spaces of interaction and exploration. In the precolonial Incan Andes, spatial representations on ceramics and through knotted strings—hardly interpreted as maps by the colonizers—focused on relative position and distinguished altitudes mainly on the basis of their bioclimatic features.[5]

In many cultures around the world, a focus on how human bodies react to changes in elevation was common: mere numerical height was generally deemed less important than the time and effort involved in a climb. This was particularly true for civilizations based in rugged areas—for instance, the Himalayas—but it occurred as well in Europe. In 1765 the French scholar Denis Diderot, famous for his *Encyclopédie*, criticized, as subjective and imprecise, attempts made to derive a formula to transform the duration of a climb into a numerical value of altitude: the time it takes to climb a mountain, he claimed, depends on multiple variables, among them the climber's speed, the route chosen for the ascent, and the slope gradient.[6]

Diderot's critique of the subjective account of altitude reflects the quantitative turn in scientific practices of the second half of the eighteenth century. Mathematical methods increasingly influenced the description of the physical world, and measurements and debates about their accuracy gained importance in scientific discourse. In 1783 the French royal engineer François Pasumot published a global synoptic table of the elevations of mountains, as ascertained by explorers and surveyors over the previous decades. On August 3, 1787, on the occasion of his famous ascent, the Swiss physicist and mountaineer Horace-Bénédicte de Saussure made the first estimate of the elevation of Mont Blanc based exclusively on differences in air pressure. He used the level of Lake Geneva as a baseline and compared his results with the trigonometric assessment of the mountain's height made by the British baronet and amateur mathematician Sir George Shuckburgh a decade earlier. At the same time Saussure noted that surrounding mountain peaks lay below the horizon, determining conclusively that Mont Blanc was the highest mountain in the Alps.[7]

Enlightenment scholars took their passion for measurement to heights around the world. Writing about his 1802 attempt to reach the summit of Chimborazo, in current Ecuador—then thought to be the highest peak in the world—the German polymath Alexander von Humboldt exemplarily combined the subjective and objective approaches to altimetry. On the one hand, he gauged elevation by recording how it affected his body; on the other, he stressed the imprecision and unreliability of altitudes defined on the basis of air pressure. He also highlighted how climbers "tend to overestimate the height they attain," then be annoyed when "confronted with correct measurements." The search for records and greater accuracy continued with the measurement of ever higher mountains and became embedded in the spirit of mountaineering throughout the nineteenth and twentieth centuries. Part of this search was the long series of endeavors to reach the highest peaks of the Himalayas, which contributed throughout the nineteenth century to a new perception of global verticality.[8]

Yet measurement of altitude required coordinated group efforts as much as heroic individual feats of exploration. Trigonometric tables, providing the ratios between a triangle's sides and angles, were developed in the third century BCE to further astronomical studies. But even before then Greek scholars—comparing the lengths of the sides of similar triangles, though not the angles—had been able to assess the height of the pyramids or the distance of ships at sea with a certain degree of precision. Few further technical developments occurred until the seventeenth century, when the Dutch mathematician Willebrord Snellius and the French astronomer Jean Picard revolutionized both the methods and the instruments used for land surveys. Primarily, they realized that to achieve greater accuracy an actual infrastructure was needed. A determination of altitude could not be the result of a single effort: measurements had to be repeated so that the rate of error could be ascertained. Triangulating large areas and comparing the heights of mountains required signals and markers that could be seen from a distance. Each definitive measurement thus needed months or years, not mere hours or days. What had essentially been the hobby of individuals now became a major undertaking requiring the kind of coordination, reiteration, and material support that could only be offered by state agencies. Accuracy and precision in establishing the heights of mountains were the products not only of technical improvements to instruments and tables, but of financial and political investments.[9]

MAKING BASELINES

Despite the evident technical differences, barometric and trigonometric measurements have one thing in common: their precision depends on the choice of reference point or, to introduce a technical term that will appear throughout the book, of *vertical datum*. This is not a preexisting, natural given, needing only to be detected in a chaotic world. Baselines for heights are invented, derived, and described rather than discovered. There is no progress to be found in the story this book tells, no constant improvement of knowledge, no approach to a more "real" system of reference. The assessment of altitude is, instead, the outcome of specific material and historical practices.

In Europe and the Mediterranean world, premodern measurements of elevation would refer, generically, to the "lowest place on earth." That might mean a local, relative marker, such as the level of water in a nearby lake or river, or a customary location, such as a church's threshold. When sea level was used, it was often just because the sea was close by. In the eighteenth century, though, it became increasingly common to relate elevations further inland to sea level. This followed on the development of the first extensive leveling networks for military purposes, the transformation of topography and geodesy into applied sciences, and steady improvements in the methods used to calculate heights from atmospheric pressure readings. To compute a meridian line, distances first needed to be reduced to a shared level, thus making them comparable. The needs of colonial administrations to conduct and compare land surveys across oceans and the growing fascination with the quantification of achievements in mountaineering made the possibility of a standardized reference framework increasingly desirable. It remained unclear, however, what exactly was meant by "the level of the sea," and surveyors rarely explicitly clarified how their zero was ascertained. The stability of the sea and its reliability as a point of reference was still a matter of debate. A variety of theories, many connected to the tale of the biblical flood, envisioned a sea that could change its level on both local and global scale.[10]

Baselines have become integral elements of our environment, part of a long list of concepts produced by humans through what we call cultural techniques.[11] As the German media theorist Bernard Siegert states:

> *Man* does not exist independently of cultural techniques of hominization; *time* does not exist independently of cultural techniques for calculating and measuring time; *space* does not exist independently of cultural techniques for surveying and administering space; and so on.[12]

Similarly, mean sea level—like other height reference points—does not exist independently of cultural techniques for the appreciation of verticality, and its changes do not exist independently of the methods used to assess rates of change over time. But once created within a specific social and cultural setting as a tool to make the world more legible, sea level becomes quintessential in shaping the environment as we know it. Human cultural conceptions of what sea level is, which individual points should be singled out from the continuous curve of tidal movements, and how absolute and relative changes can be assessed are historical constructs that have a substantial impact on how humans imagine and frame the environment.[13]

Mean sea level is thus just one of many possible benchmarks, just as the meridian running through Greenwich, England, is one among many that have been historically used as a reference for longitude. Depending on the purposes for which they are intended, different sea levels have been selected and used as zero: when the relationship to the sea is primarily defensive, concerned with preventing storm floods and the like, the main interest has been to record the highest high tide—the farthest inland point reached by the sea in its regular fluctuations. The average level of high tides is still customarily used on maps to mark coastlines—the extreme boundary of land in a strict sense. In the eighteenth century most measurements in Europe referred to high water; these figures were easier to acquire than low-tide measurements and of more immediate import to dock operations. Ports, by their nature, are not supposed to experience low tide to its full extent; no port, that is, should ever be dry. Thus, while high water can easily be measured at a port entrance, measuring low water requires a second staff some distance offshore. This does not mean, however, that low-tide levels have been ignored. When the focus is more on navigation, in fact, preeminence has been given to the lowest low tide, to indicate the minimum available depth of water and ensure that no ship runs aground when approaching the coast at any point in the tide cycle. In nautical charts some iteration of the latter has accordingly been used as the datum. The same reference point is also frequently used as the hydrographic zero, the starting point for the assessment of tidal movements, as it allows one never to use negative numbers.

The points of low and high tide can be seen or touched as water lingers at the extremes of its cycle, in what is called slack tide. In contrast, mean sea level is a pure mathematical abstraction of the tides, a "temporal average meant to smooth the variability of shorter time scales."[14] The conceptualization of sea level as an average is only one step in the long-term pursuit of a point of reference in a space that is never at rest: the littoral. As Rachel Carson writes in *The Sea around Us* (1951), "The shore has a dual nature, changing with the swing of the tides, belonging now to the land, now to the sea." There is no strict boundary, even if coastlines on maps appear precise. Neat distinctions between land and sea are relatively recent products of the modern age. As the geographer Paul Carter explains, the coastline of modern Western cartography "is an artifact of linear thinking, a binary abstraction that corresponds to nothing in nature." Coasts are actually ecotones, spaces in which different ecosystems meet and interact, porous regions that are part land, part water.

The conceptual splitting of these shifting environments into discrete elements began in the late eighteenth century. The rise of capitalism, the first wave of industrialization, and the growing infrastructural needs of nation-states made land and sea more and more into reciprocally alien worlds. Property, management, and control all require the subdivision of space into clear epistemic and legal categories. "The drawing of these lines of separation through technology and law," writes the legal historian Debjani Bhattacharyya about the differential definition of land and sea in and around Calcutta in the nineteenth century, "also entails forgetting the soaking ecologies in order to embrace the dry cultures of land use." The measurement of tides has been one of the earliest steps in understanding the relation between coast and sea, a topic of scientific inquiry of great interest for both navigation and the military. One and a half centuries of scientific research on tides led, however, to the realization that none of the haptically perceivable levels of the tide could represent an actual surface level and thus help to connect topographical surveys undertaken in different lands separated by the sea.[15]

But sea level is more than a mere boundary marker. It is increasingly also an indicator of change: a material gauge whose modifications highlight variations in the world's climate. In the introduction to the proceedings of a 2015 workshop dedicated to the integrative study of mean sea level, the French pioneer in space altimetry Anny Cazenave and her colleagues write, "Because it integrates changes in several components of the climate system in response to external forcing factors and internal climate variability, sea level is one of the best indicators of global climate change." It used to mark change on a geological timescale, evincing shifts discernable only through long-term comparisons, between, for instance, glacial and interglacial periods. These same changes are now much accelerated, showing, almost in real time, how humanity continues to transform the planet. Sea-level rise is already encroaching on places as disparate as Miami, the Netherlands, Bangladesh, and the island states of the South Pacific, prompting a multiplicity of place-specific adaptive responses to a global disaster.[16]

After millennia of unusual stability, sea level has started, in the last couple of centuries, to rise due to the impact of human activities. The rise gained enough momentum to be noticeable only in recent decades. Accordingly, the realization that a global sea-level rise is going on is fairly recent. The problem is not so much the difficulty in making sense of geological time as such. The issue is that changes that used to happen on that longer scale are now happening and affecting us on the scale of political time—a reality we are not equipped to counter or even address. Furthermore, the role of sea-level variations as agents of geological change has been subject in the last few decades to a radical qualitative transformation. In view of these recent developments, the level of the sea has morphed from an often overlooked term of comparison for geodetic measurements into a trope of climate discourse, the primary metric used to assess relative changes in the positioning of land and sea along the world's littoral spaces.[17]

THE DAWN OF DATA SERIES

Individual gaugings of sea level produce static images of a dynamic environment. Even averages based on short data series can produce distorted results due to transitory meteorological conditions, such as winds or changes in air pressure. If such series are used as baselines and applied at a larger scale—as, for example, when early cartographic work in the region of the current Suez Canal or initial efforts to set up a trigonometric survey in India adopted limited series to overcome time constraints or financial limitations—minor errors can be magnified. As

Snellius and Picard suggested, reliable height reference points can be defined only through repeated measurements that allow an averaging out of inconsistencies and determination of the rate of error. Continuity is of the essence.

The earliest available data series on sea level relative to land began to be produced consistently almost five centuries ago in Amsterdam. Elsewhere, tide gauges—the tools necessary to collect sea-level measurements—were installed in an unorderly, piecemeal fashion. Such haphazard and unbalanced development and the consequent unevenness of available series has produced biases in our historical understanding of sea-level rise. Phenomena characteristic of the North Atlantic, where most of the earliest measurements were taken, were superimposed on all the seas, under the assumption that processes such as sea-level rise would be uniform across the globe. Recent research has shown how other factors, such as the gravitational pull exerted by land and ice masses, play a crucial role in the global distribution of seawater. The loss of attractive force caused by the melting of major ice masses, such as the West Antarctic and Greenland ice sheets, may counterintuitively result in a decrease of sea levels in the immediately surrounding areas. Sea-level rise due to global glacial melt will thus affect more significantly the regions around the equator.[18]

The preeminence of the North Atlantic in historic sea-level data is, first, a peculiar product of the attempts to live and thrive in the littoral spaces of Western Europe. While the earliest known tide tables date back to China in the eleventh century CE—the product of an early interest in the wonders of the Qiantang River tidal bore—these series aimed at registering and forecasting the times of tides, rather than assessing the sea's level. Extremes in the level of the sea, such as those due to storm surges and tsunamis, have been registered for even longer: the first so-called tsunami stones, serving both as memorials and warning signs marking the upper edge of inundations, were set up in Japan as early as 869 CE, in the aftermath of the Jōgan tsunami on the Sanriku coast in the northeast of Honshū, the archipelago's largest island. But records of extremes alone provide no insights about the different levels reached by the sea over time.[19]

The idea of materially gauging ordinary states of sea level is grounded on certain muddy shores of Western Europe. For example, in Venice, the epitome of a lagoon urban environment, the blackish-green line that marks the upper limit of algal growth on the foundations of the city's buildings has served at least since 1440 as a benchmark for the depth soundings undertaken to assess the impact of siltation and sedimentation. This was a crucial effort, yielding measurements essential in determining the environmental policies of the Republic of Venice and its efforts to maintain the amphibious character of the city: it contributed both to maintaining the city's role as a trading and seafaring port and to preserving the lagoon as a defense from possible attacks from land. In subsequent centuries this natural marker, called the *comune marino* (sea average), was complemented by the carving of the letter "C" into the stone at the level marked by algae. These markers are, however, of an extremely local nature and record the level of the tide only in a single canal on a specific *palazzo*. Given the peculiar conformation of the Venetian canal network and the time it takes the tide to reach different areas, this level may vary to some extent from one corner of the city to another. As had been noted by the eighteenth century, the variability of this reference point made it impossible to assess changes over time in the relative level of the sea. A standard, mean level of high water, the *comune marino normale*, that could act as a baseline for future measurements and serve as a stable citywide datum was selected only in 1825.[20]

As mentioned above, Venice was not the only place in Europe with a vested interest in keeping tabs on the relative position of land and sea. A formal record of sea height began to be kept in Amsterdam by 1556, consequent to a court trial about the enlargement of the town walls. On the basis of these records, the locks allowing access to the new port area of Lastage would be shut when the sea surpassed a certain level, to prevent the city from being flooded. A thorough record of the highest level attained by the sea was essential to the operability and effectiveness of a port that in those years was radically improving its docks to become a node of intercontinental commerce. In 1675 the mean of these high-water levels—approximately 14 centimeters above mean sea level—was adopted as the town's official vertical benchmark, the baseline in respect to which the minimum height of dikes would be determined. In 1818, as a consequence of both the European trend toward the standardization of measuring units in the aftermath of the French Revolution and the increased infrastructural needs of the Dutch state, it was adopted by King Willem I as the standard datum for the whole of the Netherlands, with the name *Amsterdam peil*, or Amsterdam ordnance datum.[21]

TOWARD THE MEAN

Both the Venetian and the Dutch efforts were exclusively concerned with recording the average high tide, which was easily registered. Much less intuitive is the recording of both low and high tides; waves and wind constantly distort the data, making the level of the sea hard to ascertain in any given moment, even against the most accurate graduated rod. At the end of the nineteenth century the German geodesist Friedrich Robert Helmert would term this issue the "constant agitation" of the sea. To overcome this churning, the English natural philosopher Robert Moray had, in 1665, proposed the use of a stilling well connected to the sea by a channel, and thus isolated from disturbances produced by meteorological conditions. Moreover, he proposed a floating device, placed in the well, that would move a counterweight via a system of cables and pulleys. The level of the sea would then be read on a dial by means of a pointer connected to one of the pulleys. Moray was also the first to suggest that sea level should be read continuously, rather than just at tidal extremes. While the plan was never put into practice by Moray himself, it proved crucial in the scientific and technical development that would contribute to making mean sea level the standard geodetic vertical datum.[22]

Interest in accurate sea-level measurements and the creation of new technical instruments may also have been stimulated in late seventeenth-century Western Europe by the lack of actual data to test new theories about the tides. In France, for instance, the need for data to test René Descartes's theory that tides were due exclusively to the influence of the moon led the Académie royale des sciences to circulate a formal protocol about how to gauge sea levels. This effort, the first coherent venture in tidal measurements by the central administration of a state, anticipated British state-funded efforts by more than a century. It made use of stilling wells, as Moray had proposed, but radically simplified his apparatus, doing away of the complex system of pulleys in favor of a simple graduated rod connected to a float. The outcome was a sudden increase in the amount of data available about the level of the sea along the coasts of France. This technical development paved the way for the construction in the early nineteenth century of the first automated tide gauges.[23]

The potential of these scientific and technical improvements was not, however, realized. Measurements were taken discontinuously, in individual bouts separated by long periods of inactivity. And an idea prevalent among scholars during the eighteenth century, that the sea was steadily falling—based more on a few local data rather than any coordinated, comparative efforts—made it harder to adopt sea level as a benchmark for heights. Neither a single method for ascertaining an accurate sea level nor an unambiguous definition had by that time been adopted. The growing debate of those decades, though, laid the groundwork for its later definition as the standard vertical datum.

The way in which mean sea level came to be defined and spread through the world is the object of the book's first three chapters. Chapter 1 traces the history of the idea of mean sea level as the most reliable reference point for elevations. Chapter 2 analyzes how practical applications of this idea were connected to the infrastructural needs of national and colonial administrations. In chapter 3 I look into attempts to define mean sea level as a national and international standard. In the two following chapters I turn my attention to the process by which a mathematical construct devised as a vertical datum for geodetical work became a crucial baseline for the assessment of anthropogenic change. Chapter 4 explores how modern ideas of global sea-level change are grounded in early research about ancient shorelines, submerged forests, and glaciation. Chapter 5 provides an account of the further development, during the twentieth century, of theories of sea-level change induced by historical climate change and looks into the growing acceptance of the geological agency of humans. The final chapter is dedicated to present debates about sea level, including an account of the material limits of idealized averages and of the environmental histories connected to a rising sea, from Miami to Bangladesh, Tuvalu to England.[24]

Over the last two centuries of the Holocene, human ideas about sea level and how it varies have changed radically. In the eighteenth century, in continuity with older scholarly debates about the biblical flood, sea level was generally considered subject to continual decline. The earth, its mountains and coasts were shaped primarily, many scholars affirmed, by the falling, over millennia, of a proto-ocean. From the turn of the nineteenth century, the stability of the sea and its use as the standard reference point for elevations became embedded into the Western scientific canon. Finally, in the last decades, the reality of a rising sea due to the impact of human activities on the geological scale has gained wide scientific acceptance.

In this book I offer a complete view of the evolution of the concept of sea level. By connecting historical views on the sea's instability with recent material transformations along the world's coasts, I address the conventional nature of the choice of mean sea level as a stable benchmark and the renewed variability of our reference systems in the Anthropocene. I look at the framework of measurements and baselines that connects humanity and the ocean from an epistemological and conceptual point of view. In doing so, I offer a novel approach to the cross-disciplinary understanding of the environmental history of coastal ecotones and the related sciences. The world is changing, as are human ideas about it. To understand the future challenges of humanity in an environment that is less and less comparable to the one that nurtured our species over millennia, we must read the history of both oceanic sciences and surveying in a way that places them in a dialectical relationship with the environments they claim to study. New ways of doing so may be the necessary outcome of unprecedented environmental changes in the Anthropocene.

ONE

FINDING SEA LEVEL

Imagine crisscrossing all of the earth's continents with canals: the level of the water connecting the oceans through these canals would be representative of the planet's surface. This is the thought experiment the German mathematician and geodesist Friedrich Wilhelm Bessel proposed in 1837. In his youth Bessel had dreamed of taking part in overseas expeditions. To that end, he taught himself astronomy and navigation while pursuing a commercial apprenticeship in Bremen. With his thought experiment, Bessel stressed the difference between the earth's physical surface, with its elevations and depressions, and its geometrical one. While the first, he wrote, reflected the periodic variations of the tides, as well as occasional geomorphological transformations of the land, the other—which he defined as an equipotential surface that is always perpendicular to what in current terms we would call the force vector of the planet's gravitational and centrifugal forces—is characterized by a greater uniformity and stability. Since the surface of a liquid in equilibrium is by nature perpendicular to such forces, Bessel posited that the oceans offer the best approximation of the planet's geometrical shape. If, he hypothesized, there were no oceans, the choice of a vertical datum would be completely arbitrary.

But since the earth has extensive water cover, Bessel thought it appropriate to adopt it as the standard geodetical reference plane.[1]

Views on the long-term stability of the sea and its viability as a plane of reference for elevations on land had changed radically in over a century of scientific discussion ahead of Bessel's visualization. The debate was never just about the sea as such, but rather about how it relates to the land and how they shape each other. Scholars interested in understanding how mountains and coasts were formed had begun in the early eighteenth century to discuss historical and ongoing sea-level changes and to connect them with biblical accounts of a flood. The dominant theory that crystallized at that time took for granted the long-term stability of the land, with its forms gradually emerging as the level of a global proto-ocean steadily decreased.[2] The natural philosopher John Playfair, writing in the early nineteenth century, sought to explain critically the idea's success:

> The imagination naturally feels less difficulty in conceiving that an unstable fluid like the sea, which changes its level twice every day, has undergone a permanent depression in its surface, than that the land, the *terra firma* itself, has admitted of an equal elevation.[3]

In this chapter I explore the trajectory that led scientists and practitioners across Europe from theorizing a continuously dropping sea level to framing the sea as the most stable and reliable plane of reference for height measurements on land. The figures I consider include hydrographers working for various countries' naval forces, surveyors pursuing a standard benchmark for altitudes, researchers exploring the mechanics of tides, and geologists trying to ascertain how the earth's crust has changed over time. Furthermore, I show how this developing scientific understanding was intertwined with a quest to find practical ways to determine sea level. Global trends had been inferred from a few local cases. Despite being a small, shallow sea, marginal to the global oceans, the Baltic Sea played an uncannily important role in the early history of studies of changes in the relative levels of land and sea.

MAKING A STABLE SEA

Tidal movements along the shores of the Baltic Sea are minimal; that is, high and low water differ by just a few centimeters. The amplitude is slightly greater close to the Skaggerak and the Kattegat, where the Baltic connects to the North Sea, due to the influx of oceanic tides. Moving farther east and north it gradually diminishes, and most fluctuations in sea level are due to the action of the wind. All in all the Baltic appears as an utterly stable sea, whose level can be ascertained quite reliably. Despite occasional wind-induced oscillations, local seamen of the early nineteenth century claimed they could eyeball "whether the sea is two or three inches above or below its standard level." But notwithstanding the virtual absence of tides, over the centuries significant changes had been recorded: towns and ports along the sea's Swedish coast had had to be moved repeatedly as the coastline advanced and onetime islands were engulfed by the land.[4]

The Swedish physicist Anders Celsius, better known for his temperature scale, set out to quantify the anecdotal accounts of pilots, sailors, and fishermen. On repeated trips along the coast of the Gulf of Bothnia, the northern arm of the Baltic Sea stretching between Sweden and Finland, he became aware of rocks low to the water on which seals rested. These seal rocks had for centuries been the object of hunters' attention, the positions of favored hunting spots recorded in wills and otherwise passed from one generation to the next. But as the relative level of the sea dropped, the rocks jutted higher; their tops became less accessible to seals, and their value to hunters declined, a loss documented in tax papers. The property rights and financial interests of seal hunters, that is, contributed inadvertently to documenting a progressive variation in the relative position of land and sea. Celsius was thus able to compare past and present sea levels and determine a mean rate of perceived water decrease in the region over the previous century. He wanted, however, to do more. To foster *future* hydrographical study on the Swedish coast, he wished to provide a reference point researchers could use to monitor further changes. Johan Rudman, a teacher who had helped Celsius collect data about seal rocks around the city of Gävle on the southern stretch of the Gulf of Bothnia, was one of the many local collaborators involved in this quest. In 1731 Celsius asked him to carve a mark at sea level on a rock by Lövgrund island, a few miles off Gävle's port.[5]

Celsius's evidence fit nicely within the narrative of a constantly decreasing sea level that was then becoming the dominant theory. Yet there were critics. Many considered Celsius's figures to be inconsistent with the biblical account. Other scholars held that since such decreases had not been recorded elsewhere in the world, sea levels must have been essentially stable throughout history. Johannes Browallius, the bishop

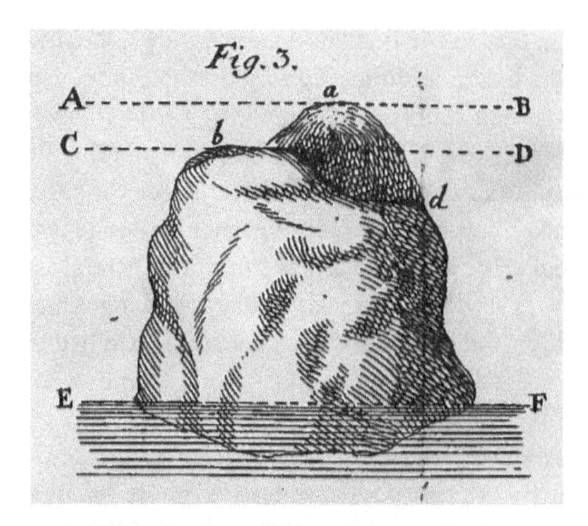

FIGURE 1.1. Based on archival sources regarding the activities of seal hunters, Anders Celsius published a drawing of a seal rock showing the changing levels reached by water over the course of the two previous centuries. *Source:* Anders Celsius, "Anmärkning om vatnets förminskande," plate II.

of Åbo/Turku, for instance, thought that the apparent decline Celsius reported was probably due to measurement errors. To bolster Celsius's theory by definitively proving the trend, Anders Wijkström, a professor of mathematics based in Kalmar in southeastern Sweden, began a series of sea-level gaugings. On May 21, 1754, he placed a scaled gauge on the city's seaside walls and began recording daily observations. In August 1755 he had a T-shaped mark carved on a rock of Skällo island, a mile off the port of Kalmar. The idea was to analyze changes in sea level over time by comparing the annual average of gaugings taken at the port with the baseline recorded by the mark at Skällo. The method proved particularly valuable, as it not only tracked the overall trend but recorded the rate of change and its yearly variations. Extending Wijkström's work, new data were collected more than forty years later in the same location by the professor of mathematics Anders Frigelius.[6]

The debate about Celsius's results did not remain confined to the Baltic region. Around the time Frigelius made his new run of measurements in Kalmar, Playfair, in the notes to his popular account of recent advances in geological theory, dismissed all of the previous century's theories of sea-level change as "guided more by fancy than reason." Any local fluctuation in the absolute level of the sea would need, he claimed, an equivalent change all over the earth, and loss or accretion of water on a planetary scale would require some complex causal mechanism. Local variations in land elevation appeared to him to overcome the need to devise such a mechanism, and might help to explain the divergent trends recorded in different seas: the Baltic falling, the North Sea and Mediterranean allegedly rising, and some seas in the tropics, again, decreasing. The geographical inconsistency of the observations allowed him to dismiss other theories as well, such as ones suggesting that ocean levels were rising toward the equator because of changes in the planet's rotational velocity. Playfair's idea of a stable sea and changing land greatly reduced the need for such complex theories, offering a more elegant way to explain recorded local and regional differences and changes over time.[7]

A few years later the German geologist Leopold von Buch proposed his own "theory of elevation," according to which the perceived change on the Baltic coast was caused by rising land. Originally he had been a supporter of the idea of a falling sea, but faced with increasing amounts of discordant data, he reversed his beliefs, stating in 1810, "It is certain that the surface of the ocean cannot *sink*; this the equilibrium of the sea will not by any means admit of." This claim for the sea's stability at first met with disbelief in the scientific community.[8] In the explanatory footnotes he wrote for the English translation of Buch's work, the Scottish mineralogist Robert Jameson tried to minimize the consequences of the new theory, claiming that Buch actually referred to a movement of the sea even when he clearly refuted it. Jameson quoted the writing of another German scholar, the uniformitarian geologist Ernst Friedrich Wrede, to help smooth over the dispute:

> M. Wrede of Berlin, in his "Geognostische Untersuchungen über die Sudbaltischen Länder," conjectures the centre of the globe to have undergone a slow change, being transported along the axis from the *north* to the *south*, and hence that gradual *depression* of the level of the sea in our hemisphere, which he supposes to be accompanied by its gradual *elevation* in the southern hemisphere. It would appear that M. Von Buch had in view this notion of Wrede's, when he considered it as highly probable, that the land in the northern regions was *slowly rising above the surface of the sea.*[9]

In part due to the influential biblical account of receding water after the flood, the idea of a generalized sinking of the sea kept its theoretical predominance for another few years. As for the Baltic Sea, in 1820 and 1821 the Swedish navigation service collated a series of measurements of sea-level changes over the previous forty years, establishing many new watermarks. The service's pilots had a working mastery of specific stretches of the coast and detailed knowledge of water depths and their variations. Joining their local expertise in a national project allowed scientists to greatly expand the geographical extent of Celsius's observations. It thus became possible to establish that changes in the relative position of sea and land occurred at different rates in different parts of the sea. While this would have been more easily explained by local movements of the earth's crust, the Swedish cartographer Carl Petter Hällström justified Celsius's preference for a falling sea, claiming that all discrepancies in local rates of decrease were due to peculiar geographical and meteorological conditions.[10]

The theory of a falling sea lost favor, however, in the following decade. In *Principles of Geology*, first published in 1830, the Scottish geologist Charles Lyell used examples like the Baltic observations to strengthen the case for local land uplift and against actual fluctuations in sea level as the cause for the recorded variations in the relative position of land and sea. The most prominent other example discussed by Lyell was that of the Temple of Serapis at Pozzuoli in southern Italy, which had been submerged by the sea multiple times since the Romans. In the face of the overwhelming evidence presented by Lyell, the dominant theory favoring a steadily decreasing sea lost support, and the idea of a stable sea became increasingly widespread.[11]

By 1834 George Bellas Greenough, then president of the British Geological Society, publicly withdrew his own diluvialist theories of fourteen years prior. Ever the pragmatist, he warned, however, against rapid acceptance of any of the numerous causal theories for land uplift or crustal movements as the sole explanatory device for relative changes in the position of land and sea. Nonetheless, Lyell's idea of a substantially stable sea facing a moving crust became the dominant theory throughout the remainder of the nineteenth century.[12]

The scientific dispute about relative sea-level change on the Swedish coast of the Baltic Sea exemplifies certain knowledge-production processes. It played a crucial role both when the dominant theory held that land-formation was a product of a falling sea and when the interaction of a stable sea with moving land became the favored explanation for changes in sea level. In both cases, scientific consensus was the product of limited knowledge about local environmental conditions giving shape to a universal theory. The conformation and peculiar geological processes of the Baltic created, thus, the preconditions for a lively scientific debate that, over the course of a century, revolutionized the understanding of sea-level variations.

STATE OF EQUILIBRIUM

Growing consensus around the idea of a stable sea strengthened the viability of sea level as the reference point for elevations and coincided with efforts to exactly define mean sea level. While sea level was adopted by surveyors with increasing frequency through the eighteenth century, in most cases little if anything was said about how a level that represented the sea in a state of equilibrium could be determined in less favorable conditions than the Baltic Sea. Indeed, "mean sea level" was applied liberally to any sea level used as a reference plane, regardless of how it was ascertained. In early nineteenth-century Britain, for instance, the idea predominated that low water represented a surface level—that is, a level that, at a certain moment in time, is the same at all points along the coast. This was primarily due to the role of navigation in the country's economy; low water was already a common benchmark for depth soundings. Moreover, it minimized surveyors' need to use negative numbers when making measurements in the littoral zone. When in 1837 a datum had to be selected for Ireland, this rationale led surveyors to select low water at spring tide at Poolbeg Lighthouse near Dublin.[13]

Many attempts to estimate the level of the sea at rest ended up just averaging high and low tides. But not all theorists were convinced that this approximated the "natural level of waters." In 1781, for instance, the French astronomer Jérôme Lalande suggested that the natural level of the sea was instead to be found nearer low tide, at about a third of the tidal range, because of the almost elliptical shape of the planet. This interpretation derived from a broader debate on the causes and effects of the tides, but it never caught on among practitioners looking for a reference plane. Then in 1798, the mathematician Pierre Simon Laplace discussed the equilibrium state of the sea as the level it would take, at any given moment, in the absence of solar or lunar influences—an understanding that is close, if not essentially identical, to later definitions. Laplace still understood the level of water more as an abstraction than as a physically determined reference point, and was more interested in establishing it through astronomical theorization and calculations than by gathering and comparing actual sea-level data over time.[14]

Lalande and Laplace did not, however, stop at providing conceptual definitions of the ideal level of the sea: they acted practically to further its ascertainment in the port of Brest, the main base of the French navy on the Atlantic. Tides had been recorded there since 1679, repeatedly but not continuously—always for just a few days at a time. While these scattered observations made it possible to reconstruct rough trends in the tides, they lacked the continuity needed to determine mean sea level and assess its variations. In 1800, Laplace wrote to the maritime prefect of Brest—an office recently created by Napoleon to unify the local command of harbors and the navy—urging him to establish regular measurements from which to derive the mean level of the sea for use as the gauge's zero. Similar requests had been made around the same time by Lalande. In 1803, Laplace, as part of a commission charged with improving tidal observations along the coasts of France, elaborated the suggestion in a published report. Together with the hydrographer Pierre Lévêque and the Brest-based astronomer Alexis de Rochon, he encouraged the installation of an observatory in the port to produce regular, ongoing assessments of the tides. Possibly in response to this report, a continuous series of measurements of the times and heights of low and high tides in the port of Brest was finally established in 1807.[15]

The increase in data, combined with advances in geodetic theory-building, such as Bessel's thought experiment about canals, made it more and more common to use some iteration of sea level as a reference point. If the sea was not only stable and uniformly spread over the planet, but also representative of the planet's geometrical surface, what better way to compare heights across stretches of sea or at great distances. The level of the sea could be assumed to be zero anywhere it was measured, saving the cost and effort of long surveys across land. This conclusion, however, still required a way to calculate or materially assess the state of equilibrium of the sea.

THE RISE OF HALF-TIDE

After the fall of Napoleon, the hydrographic office of the French navy, led by Charles-François Beautemps-Beaupré, was put in charge of charting the country's shoreline. Peacetime seemed an opportunity to satisfy a long-felt need for better coastal maps. All the existing ones were faulty in one way or another, and the navy lacked the data to fix them. Focused on aiding navigation, the hydrographic office prioritized the recording of tidal extremes, in particular lowest low tide, then used this information to standardize depth soundings along the coast, a crucial step in producing usable navigation maps. While fixated on this task, the effort, along with coeval processes in other countries aimed at improving knowledge of ocean spaces, produced a vast amount of data about the whole range of tidal movements in multiple coastal localities. Still, though, measurements were often done over relatively short time spans, sometimes just a few hours.[16]

Following Laplace's lead, French surveyors took on the challenge of comparing seas in "a state of absolute rest." By the mid-1820s French army topographers repeatedly claimed to be using the mean level of the sea, or *mer moyenne*, as the standard reference plane for their mapping efforts. In 1831, Pierre Daussy, an assistant curator at the navy's cartographic office, read a memoir at the Academy of Sciences in Paris on the measurement of tides along the coasts of France, in which he described the method used to calculate the mean level of the sea as pretty obvious: "In fact, whenever two high tides and one low tide, or two low tides and one high tide, have been observed on the same day this mean level could be obtained by taking the half-sum of the intermediate tide and the average of the two extreme tides." Rather confusingly, intermediate and extreme here do not refer to the tide itself but to the chronological order in which the tides were observed. In the first case described by Daussy, the intermediate tide would thus be the low, while the extremes would be the two high tides.[17]

A characteristic example of the efforts by French topographers to calculate the mean level of the sea was the work of the army topographer Jean-Baptiste Corabœuf between 1825 and 1827, when he was in charge of surveying the Pyrenees. At the core of his work he put the question of whether the Atlantic Ocean and the Mediterranean Sea were part of the same surface level. In an appendix to his report, Corabœuf explained how he determined the mean level of the sea at the geographical endpoints of his survey. On the Mediterranean side, he was dissatisfied with the datum then in use: the level of water in the Étang de Leucate lagoon near Perpignan, which was connected to the sea only by a narrow passage and thus less affected by tidal movements than the open sea. The accuracy of this traditional method of estimating the mean level of the sea had been questioned by some surveyors, and Corabœuf decided to make his own measurements. To do so, he set up a simple leveling staff fifteen meters offshore by Fort St.-Ange in Perpignan and ascertained the variations in sea level over about three hours on two separate occasions at different times of the day. He then averaged the lowest and highest levels of water recorded in the two sessions and used the resulting level to determine the height of a nearby elevation, which he planned to use as his main benchmark. Having established that the two mean levels differed by only about a centimeter, he decided to adopt their average as the starting point for his work. When he compared his results with the earlier estimates based on the level of water in the Étang de Leucate, Coraboeuf found, much to his surprise, that he had confirmed the latter's accuracy. At the Atlantic end of his surveying chain, Coraboeuf adopted as zero the mean level of the sea as computed by the navy hydrographers coordinated by Beautemps-Beaupré and Daussy on the basis of thirty-nine days of data gathering at Socao Fort, close to the border with Spain.[18]

At about the same time in Britain, "half-tide"—as the mean tide level thus determined has been called—was being adopted by an increasing number of observers as a rough approximation of the level of the sea at rest. There as elsewhere, infrastructural ventures—such as the consolidation of a canal network and the birth of railways—necessitated a clearer definition of which sea level made for a good reference plane. At the end of the eighteenth century, William Mudge, the director of a trigonometric survey of England and Wales, had referred to the sea to assess altitudes—but did not specify how its level had been ascertained. In 1823, the Board of Longitudes—broadening its interests once its original question, how to determine longitude, had been satisfactorily solved—set up a tide gauge on the Thames by Greenwich. The sea level measured at this gauge was supposed to act as a standard of reference for the level of the sea. While (as in the French case) no details are available about the exact purpose and characteristics of such a standard, the fact that investing time and money to establish it was deemed worthwhile is itself interesting.[19]

In 1830 the globetrotting surveyor John Augustus Lloyd was commissioned by the Admiralty to chart the differences in the level of water at particular locations along the Thames, starting from the Royal Dock in Sheerness, on the river's estuary. Using observations made from 1827 to 1829, he calculated the mean of the tides in the dockyard. The observations were incomplete and discontinuous, which spurred him not only to calculate the overall average but to compare it with the mean for 1827, the year with the most data points in the series. Finding the results of the two runs of calculation to be extremely close, he adopted the mean for the whole three years as the reference point for his survey. This was probably the datum based on the longest series to date. Nonetheless, these observations of high and low water had been made irregularly, and the results were limited in their accuracy. To promote the production of continuous and more accurate measurements, the Admiralty funded the installation of a new tide gauge, designed by Lloyd himself.[20]

AUTOMATING MEASUREMENT

The installation around 1830 of the first self-registering tide gauges, in David Cartwright's words, "changed the very concept of tidal variation from a daily sequence of extrema to a continuous process in time." Continuous data opened the way to computing mean sea level as it is now understood and was central to making its adoption as the standard point zero of elevation a practical possibility.[21]

A self-registering gauge afforded surveyors and other practitioners the possibility of precision all but unattainable previously. With a traditional gauge, shifts of at least two officers would have needed, with the help of a chronometer, to register the height of the tide at regular intervals—day and night, week after week, maintaining a precise routine through darkness, storms, sloshing water, and creeping cold. To record just tidal extremes accurately, suggested the British astronomer royal George Biddell Airy in the 1840s, required readings at five-minute intervals starting "decidedly" before and ending well after high and low water. And the same procedure would have to be repeated with the same regularity at multiple locations to register differences between local gaugings.

The limits of fieldwork as a mode of knowledge production have indeed added to the intrinsic imprecision of all measurements of the level of the sea. In his 1829 account of the survey of France's western coast, Beautemps-Beaupré remarked how hard it was to find people devoted enough to the needs of the hydrographic service to spend six months alone in a small coastal village observing every change in sea level with impeccable regularity. Yet harder, he quipped, was convincing them not to make spurious log entries when they forgot or were unable to take a measurement. And even those rare officers who could be trusted completely were not immune to error, its causes ranging from the displacement of a gauge or failure of a clock to mere bad weather.[22]

It was precisely because nobody would attend to the tide gauge Lloyd had installed at the Royal Docks in Sheerness that Admiralty engineer James Mitchell started work on a machine to reduce the human labor required for data gathering. What was possibly the first working self-registering tide gauge was installed in 1831 in the Sheerness dockyard. The device was simple in principle, building on Moray's design: a float in a stilling well moves up and down according to variations in the water level; through an attached rod the level can then be read against a fixed scale. Mitchell's innovation was the automation of the registering apparatus: a system of wheels and gears translates the float's vertical movement to the horizontal plane, where a pencil traces a line along a graduated sheet of paper rolled around a cylinder driven by a clock. The height of the water in the stilling well could thus be registered continuously, with no human intervention except to regularly change the paper roll.[23]

Mitchell was not working in a vacuum. In the early decades of the nineteenth century the British polymath Thomas Young had played a pivotal role in reviving interest in tides as an object of scientific inquiry, proposing in 1813 that the focus on tidal extremes might at last be overcome by continuous observations of the tides' fluctuations. The idea that automation could overcome some limitations of manual observation, while improving continuity and accuracy, was becoming increasingly common. And others were, in fact, working on such projects. In March 1831 John William Lubbock, a disciple of the Cambridge polymath William Whewell, presented to the Royal Society another automated tide gauge, developed by Henry Palmer. As the engineer at London Docks, Palmer was concerned with the impact of the planned removal of London Bridge on tidal movements in the port of London. In the years prior he had begun to collect data at London Docks at fifteen-minute intervals. Palmer plotted the readings on graduated paper, gathering consecutive days of sea-level fluctuations for comparison on one sheet. The iconic overlapping curves of tidal recordings that Mitchell's apparatus would make standard were thus anticipated by hand-drawn curves. While a model of Palmer's "graphical registrer" was probably built, it was never installed.[24]

Gauges inspired by Mitchell's and Palmer's designs quickly spread. The Admiralty spent a hundred pounds each to install more of Mitchell's machines in Portsmouth and Plymouth. Already in 1833 the Admiralty used observations on the time of high water gathered by these automated gauges, in addition to data collected manually at London Docks, as the basis for the predictions printed in the first volume of its *Tide Tables*, which informed sailors about when approach to port was safe. The diffusion of automated tide gauges in turn fostered the development of further

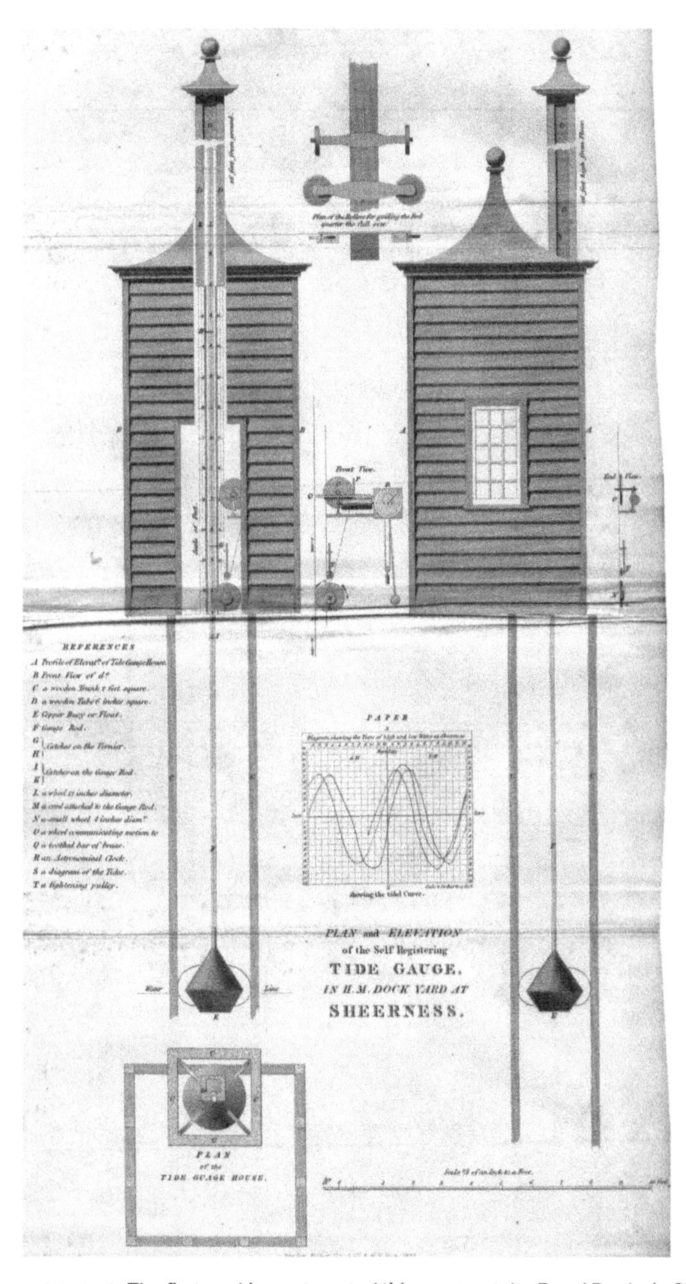

FIGURE 1.2. The first working automated tide gauge at the Royal Docks in Sheerness. The automation of tide measurements revolutionized not only the gathering of data about tidal movements but the way the mean level of the sea was conceptualized and calculated. *Source:* "The Tide Gauge at Sheerness." Courtesy of Cambridge University Press.

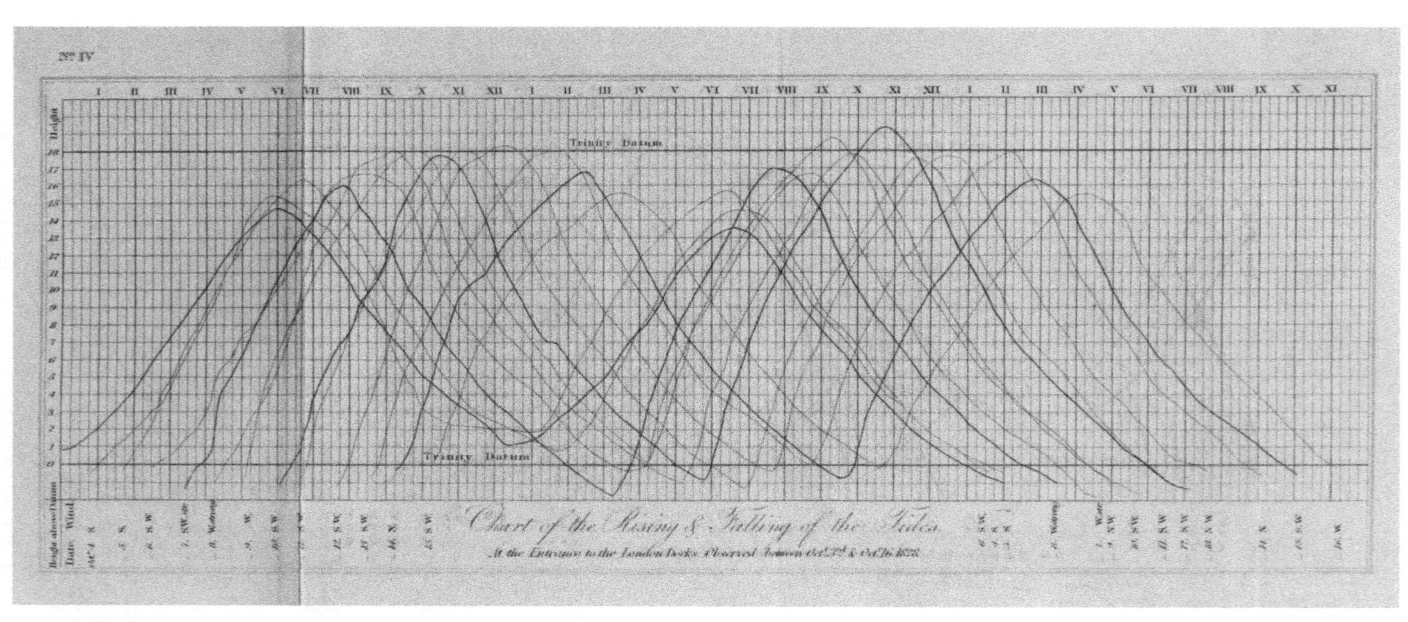

FIGURE 1.3. Even before the installation of automated tide gauges, the fluctuations of tides had been represented as curves. Henry Palmer, for instance, drew by hand curves interpolating data recorded manually at intervals of fifteen minutes at London Docks. *Source:* Palmer and Lubbock, “Description of a Graphical Registrer of Tides and Winds,” plate 4. Courtesy of the Royal Society of London.

designs and technical improvements, which soon spread abroad. France set up self-registering tide gauges in Le Havre in 1834 and Algiers in 1843; one was erected in Sitka, Alaska (then still Russian territory), in 1841; and in 1846 the first automated gauge in the United States was set up in Fort Morgan on Mobile Bay in Alabama.[25]

COMPARING TIDES

Automated tide gauges supplied more plentiful and accurate data for more and more coastal locations. But measures of low water and high tide at disparate sites remained variable and difficult to compare. Spurred by the need for a stable reference point, the new British Association for the Advancement of Science (BAAS)—established in 1831 as a less elitist response to the Royal Society, more attentive to the needs of the country's provincial centers—began to sponsor research on formalizing a standard definition of mean sea level. William Whewell, who had been drawn into the subject of tides by his former student Lubbock, would eventually lead the quest.[26]

In 1834 the BAAS set up a committee charged with ascertaining the relative level of different locations along the coasts of Great Britain. Hindered by the geographical dispersion of its members, it got off to a slow start. Nonetheless, a first result was presented in 1835, when at the BAAS meeting in Dublin, the navy hydrographer Henry Mangles Denham discussed the long-term stability of mean tide and its viability as a baseline for altitudes, a conclusion based on his survey of the rivers Mersey and Dee and the port of Liverpool. Denham suggested that a dated "half tide level" be engraved in every harbor, to serve as a scientifically sound standard of reference for future measurements. Once Whewell took charge, in 1836, he broadened the committee's efforts beyond the measurement of sea level in individual location, refocusing them on the idea "that the exact determination of three points considerably distant from each other on the coasts of this island might throw light upon several important questions"—including the real meaning of the expression "level of the sea." The committee decided to undertake a combined research effort to produce a baseline for the diachronic comparison of crustal movements along the British coast while also checking the viability of half-tide as a proxy for sea level. The aim was to get around both the impossibility of assessing shifts in the earth's crust in a single location and persistent doubts about the arbitrariness of mean tide level, that is, the average of low and high water, as a reliable datum. Once "mean water" was determined to be "at the same level at different and distant points of the coast," its usability as a reference plane would necessarily be acknowledged.[27]

The trigonometric survey took place between May and October 1837, while sea-level measurements were taken around the end of that year. The Bristol-based surveyor Thomas Bunt focused on a line passing through Somerset and Devon—a region chosen for allowing the shortest route between two separate seas—from Watchet on the Bristol Channel to Axmouth on the English Channel. Once the north-south survey was completed, it was proposed to connect it at Bridgwater with another one reaching an undefined third point in the east, possibly to record sea level on the North Sea as well. Due to the hilly terrain, surveying an east-west line proved more difficult and costly than expected; doubts also arose that, with the need to compensate for multiple changes in elevation, accuracy would be negatively affected. To make up for the inability to reach another sea, but still provide for the possibility of mearuring whether sea level varies along the east-west axis, in mid-1838 Bunt undertook a further survey along the coast of the Bristol Channel, from Bridgwater to Portishead. Tide observations were made at Axmouth, Watchet, and Portishead, for a month at each location—long enough to ensure that both spring and neap tides were included. Bunt determined that, while the levels of the Bristol and the English Channels, measured respectively at Watchet and Axmouth, were substantially identical, sea level in Portishead apparently lay more than 4 inches lower. It was suspected that this was due to the tidal observations in Portishead having been made in a different season than those in the other two locations, and in July 1838 new ones were made concurrently in Portishead and Axmouth. These yielded an even greater measured difference of 9 inches. After discussing possible causes for the discrepancy, such as irregularities in the tidal cycles and the role of sediment at Axmouth, Whewell suggested ignoring it. Since recorded differences between the tidal extremes at all three locations were much bigger than the discrepancy recorded for mean water, mean tide level appeared to be the best possible approximation of the actual level of the sea. In the 1839 report on Bunt's survey, Whewell drew also explicit connections to ongoing theoretical developments in geology, among them the debate about the stability of the planet's crust and the consequent rise of the new consensus on a stable sea.[28]

Beyond leading the BAAS effort to ascertain the level of the sea around Britain, Whewell became, through the 1830s, a leading figure in tidology, the study of the causes and mechanics of tides. This endeavor also strengthened the case for the use of mean tide level as a baseline for elevations. As part of his work on cotidal lines—that is, lines connecting ports at which high tide occurs at the same time—he amassed observations from all over the world and noticed the long-term stability of the mean in respect to both low and high water. In particular, he observed how, while consecutive low waters at Singapore could differ by several feet, the long-term mean oscillated by only a few inches. Consequently, he criticized surveyors' widespread use of low water as a reference point, arguing that it could lead to significant material error. Whewell produced a clear case for mean tide level as the most constant level of the sea over time, especially considering the major differences in the height of tides in adjoining locations: "A *level surface* drawn from any point (that is a surface of *stagnant water*) would probably be nearly parallel to the points of mean water at different places."[29]

Extending this work, Whewell also looked at the variability of mean tide level over time at a single location. Using data collected in Plymouth by William Walker, the assistant master of the local royal docks, Whewell determined that between 1833 and 1838 mean annual tide level had never fluctuated by more than 3 inches. Variability of the mean from one fortnight to the next was, in contrast, substantially greater. Whewell attributed this to the impact of local circumstances, specifically the fact that low water varied with mean lunar declination and by a greater extent than high water. On the basis of the year-to-year comparison, he reiterated that, in the long term, mean tide level is constant. To strengthen the case for the "permanency" of mean tide level, Whewell also provided a series tracing a single year of observations made at the other end of the country, in Dundee in Scotland. These observations, made available by the hydrographer Joseph Foss Dessiou, who directed the production of the Admiralty tide tables at the British hydrographic office, showed that the fortnightly inequality of mean water was confined within a couple of inches and thus that this variability was due mainly to differences in geomorphological conditions.[30]

As noted by historian Michael Reidy, Whewell's work on tides would not have been possible without the contributions of multiple collaborators. These included dock engineers like Walker, surveyors like Bunt, and computers at the hydrographic office like Dessiou. Whewell, while availing himself of their work, regarded them as subordinates and himself as the one offering a theoretical framework that explicated the relation between individual observations. But beyond doing work "integral to Whewell's tidal analysis," these other, independent actors were actively involved in redefining how the level of the sea was understood and calculated. Enmeshed in their own systems of relationships, they offered their own takes on broader theoretical issues. Walker, for instance, authored an 1846 essay on tides that placed the measurements he had made in Plymouth within the long history of developing theories about tidal movements. He explicitly presented himself as a Whewellian scientist, capable of taking observations made by others and detecting yet-unnoticed connections. The stability of half-tide around the Isles of Scilly could be readily observed in the tide tables drawn up by marine surveyor Graeme Spence at the end of the eighteenth century. Walker remarked, however, that he was the first actually to do so, and thus to place Spence's tables in the context of ongoing scientific debates about the relevance of mean tide level as a prime marker of a stable sea. Aware also of the international debate about ascertaining sea level, Walker made direct references to foreign scholars and practitioners such as Lalande and Daussy, noting their contributions to, respectively, the theory and practice of tidology.[31]

The most extensive test of the idea that mean tide level represents the state of the sea at rest was undertaken in 1842 as part of the survey of Ireland led by Thomas Colby, then director of the Ordnance Survey, the British mapping agency. The royal astronomer George Biddell Airy was involved from the beginning in this endeavor, which aimed explicitly at determining a reference plane. Airy not only advised on setting up the experiment but oversaw data analysis. Observing stations were selected with extreme care, in particular making sure to cover both the western and eastern coasts of the island, and thus detect any difference in level between an open and a closed sea. Where officers took measurements manually, efforts were made to ensure coverage comparable with that of an automated gauge. Rather than focusing exclusively on daily extremes, at least one full tidal cycle—from one high or low water to the next—was to be recorded at close intervals, so that a complete analysis could be made of the tide.[32]

Mean sea level has a conceptual history that far predates the 1830s. While formalization of mean tide level as the main reference point for altimetry is frequently attributed to Whewell, it was actually the outcome of a much longer, more intricate scientific debate. Nonetheless, two developments in the early decades of the nineteenth century radically changed the way sea level was computed and understood: the idea that the sea was geologically stable in the long term and the introduction of automated gauges.

The idea of a stable sea emerged from a debate that flourished in the eighteenth century over why certain Swedish towns, in order to preserve their ports, had been forced to relocate closer to a sea perceived as receding. The idea that the sea could actually fall gradually lost favor, giving way to the idea that sea level was stable, and thus an ideal benchmark for elevation. By late in the century, discussions of a rather generically defined "mean sea" had become increasingly common.

In the mid-nineteenth century, improvements in data collection allowed scientists and practitioners alike to revise mean tide level (the simple average of mean high water and mean low water over a certain period of time) into a closer approximation of the state of the sea at rest: mean sea level—that is, the mean of the whole tidal curve, or at least of multiple regular samplings. Due to the complexity of the tidal curve, mean tide level and mean sea level are not exactly the same. In most locations the differences are minimal, but they can rise to some centimeters under certain conditions. The possibility of a significant discrepancy was noted, in fact, by Airy in 1842 on the basis of Colby's survey of Ireland. While it was possible to manually gather enough observations of the level of the sea to calculate its mean, as exemplified by that Irish survey, the introduction of self-registering gauges made the adoption of mean sea level as the vertical datum more immediately practical and less cumbersome.[33]

Among the possible measures of sea level, its mean has a peculiarity: in contrast to other levels still used at the turn of the nineteenth century as reference planes, mean sea level cannot, in most instances, be experienced sensorily. While low, lowest low, high, and highest high tide tend to linger, becoming almost tangible, mean sea level is a product of abstraction and computation. If you stand at the shore, you can perceive the moments at which the tide turns; the level computed as mean sea, however, slides right past you. Defining the reference point for elevations as an average is an attempt at setting a standard through statistical means, at creating a virtually fixed boundary between land and sea and between water and air. In this sense, the conceptual history of the level of the sea is part of a broader movement toward standardization that characterized the nineteenth century. This was one of the central features in the development of modern nation-states, driven by the need to make the world more legible as a means to increase control of both territories and populations.

States have played a central role in the making of standards. A number of studies have traced, for example, their engagement in determining longitude at sea and in the adoption of time zones. The rise since the eighteenth century of the so-called infrastructure state, responsible for expansion and maintenance of transport, and the growing interconnectedness of the world during the nineteenth century, were paramount—as will be shown in the next chapter—in producing such developments, including an increased attention to the definition of uniform vertical datums.[34]

TWO

INFRASTRUCTURES OF MEASURE

A couple of decades ago, in the early 2000s, the construction of a new bridge over the river Rhine, across the Swiss-German border, was briefly held up by the realization that the bridge would reach the German bank more than half a meter too low. The national datums of Germany and Switzerland differ by 27 centimeters. While this was accounted for in the plans, at some point somebody apparently turned a minus into a plus, doubling instead of compensating for the difference. The issue was swiftly resolved, by lowering the access road on the German side. But one sees how, even when complicating facts are well known, the risk of error increases when multiple frameworks of reference meet. Imagine what errors can occur when the discrepancies between reference planes are not clearly documented.[1]

Some differences in standards have administrative origins, changing as one crosses a national or regional border. Others arise from the choices of individual actors with diverging agendas and needs. Before a standard is selected, two or more datums may coexist, as was the case in Belgium before mean sea-level was finally selected in 1889. For decades Belgian civil and military authorities had used two distinct iterations of mean low tide at the port of Ostende as starting points for their respective surveying efforts. When joint works were planned, the difference of almost 17 centimeters had to be considered and compensated for to avoid errors. The choice of a common reference point does indeed simplify construction projects, but the standardization of measurement frameworks has in itself been a necessary infrastructure for public works in both national and colonial contexts.[2]

The infrastructure undergirding a mean sea-level datum was developed mainly within individual national territories or as part of colonial endeavors. Some preliminary work toward the creation of a worldwide network of sea-level gauges was undertaken, alas rather unsuccessfully, by scientists and practitioners starting in the early decades of the nineteenth century. In the end, however, its realization depended on the choices of nation-states and colonial powers. Nonetheless, data collected at a growing number of locations throughout the world were interpreted by surveyors within a broader debate over the possibility that the oceans were at different levels, as many mariners believed. This theoretical debate soon found a practical application. The issue of whether oceans are at the same level affected plans made throughout the nineteenth century to create shortcuts between the world's oceans and speed up commerce networks by cutting canals through the Suez and Panama isthmuses. That early measurements seemed to support this idea, that sea level on one side of an isthmus differed markedly from that on the other, raised questions about whether such canals were feasible.

The interests of commerce played a crucial role, too, in individual countries' selection of reference planes for heights. The growing need for transport networks—for planning and building on a national scale—called for improved standardization of national surveying frameworks. With a great many interested actors, politics played as big a role as scientific insight and reliable measurement in the selection of the earliest national datums. At times politics proved more important than accuracy, as can be seen by comparing the processes in France and Great Britain. In France, the desire to settle on a stable benchmark as soon as possible pushed the government to pick a national datum even if its accuracy was questionable. In Britain, the datum was constantly revised by the Ordnance Survey according to changes in measurement practices. In both cases, it is evident how, while political priorities may differ—stability versus accuracy—datums are essentially products of convention.

Moving from the metropole to the European colonies, specifically British India, one sees how the choice of where to install tide gauges was, like the choice of a starting point for a measurement framework, the product of specific interests. A gauge might be placed, for instance, to establish a reference point for land surveys or in service of port operations and navigation. In the first case, the aim could be to approximate the conditions of the open sea, so as to claim that the measured level was representative of the overall, ideal global sea level. The second case, by contrast, prioritized actual local sea levels and the mechanics of the tide in a specific place—factors that influenced day-to-day port activities. Different sets of economic and scientific interests, supported by varying groups of actors, thus guided the process by which locations for instruments were selected.

IDEAS OF GLOBALIZATION

By the early 1840s automated tide gauges had already been installed in a handful of territories outside Europe, such as Algeria and Alaska. The first self-registering tide gauge in the United States was installed in Alabama in 1846, and a new model, developed for the US Coast Survey by renowned instrument maker Joseph Saxton, in San Francisco around 1854. In Australia the first such gauge was installed in 1858 at Williamstown, Victoria. Other locations, even within Europe, lagged behind. The first automated tide gauge in the German states, for instance, was erected in Hamburg only in 1861. Except for some European and North American gauges, most early stations did not produce continuous series of data after their installation. Most were installed with specific goals and, mainly due to financial reasons, remained in use only for as long as was deemed necessary for the task at hand.[3]

The global network of sea-level measurements developed on the backs of state-building and colonial projects. Comparing sea levels on a global scale was never the central issue. Other, local interests prevailed. This was not necessarily the way it had to be. A longtime promoter of transnational intellectual networks, the German geographer and naturalist Alexander von Humboldt stood at the forefront of the comparative, global study of sea levels. If things had developed according to his plans, he would also have been the initiator of a carefully planned network of markers of the level of the sea.

In 1829, the year he would turn sixty, Humboldt was finally able, thanks to the financial and logistic support of tsar Nicholas I and his finance minister, German-born Georg von Cancrin, to undertake a second great expedition after his famous American travels: a tour of the Russian empire, across the Urals, to the border with China and Mongolia, and, on the way back, through the Kazakh steppes, to the northern coast of the Caspian Sea. The official aim of Humboldt's first trip to Asia was to help the Russian government determine what valuable minerals could be found in its territories. True to character, Humboldt filled the months with inquiries in a multiplicity of other fields. On the way back he was expected in Moscow and St. Petersburg to report on his travels. In both cities he was greeted with interest and affection by the Russian elites. Interminable invitations to parties and receptions followed; the experience was tiresome for Humboldt, who had much preferred being on the road. The only bright spot was the speech he gave on November 16 to fellow scientists at the Imperial Academy of Sciences in St. Petersburg. High-spirited after so long in the field, he made various proposals to embed Russian science into international collaborative networks, from the development of geomagnetic studies to the collection of climate data.

Consistent with his long-held interest in isometric lines—which connect all points on a map that share a particular value—as tools of scientific inquiry, Humboldt insisted on the need to determine the Aralo-Caspian "curve of zero height"—a line linking points at the same level as the Black Sea and, consequently, delimiting the depression that surrounds the Aral and Caspian seas. He also suggested fixing markers along the Caspian coastline, creating a baseline for the assessment of future sea-level variations. If identical changes were registered at all stations, one could assume that the level of the sea itself was changing. If shifts diverged, the perceived changes were probably due to crustal movement, as Buch and Playfair had suggested for Scandinavia. The Russian scientific community welcomed the suggestion enthusiastically: the following year, after careful consideration of local geomorphological conditions, the physicist Emil Lenz placed a marker of sea level by Baku's fortress. More than a decade later, Humboldt remarked on the importance of continuing such efforts and broadening their spatial and temporal scope. To attain the most precise results, he suggested, water levels on both sides of the sea should be regularly compared at thirty-year intervals.[4]

Building on this idea, which aimed at understanding the natural oscillations of a closed sea, Humboldt asked polar explorer James Clark Ross to place similar markers during his expedition in the southern hemisphere, to help answer more general questions about the physics of the globe and its oceans. Accordingly, in 1841 Ross engraved a permanent marker of the level of the sea in Port Arthur, Tasmania. While Ross reported—either in error or to appease Humboldt—that the marker recorded mean sea level, modern estimates suggest it was actually installed about an hour before high water. Indeed, the Port Arthur sea level was probably just eyeballed, a rough approximation of mean sea level as recorded by the manual tide gauge installed a few years earlier by T. J. Lempriere, a local amateur scientist and storekeeper. The stated aim of Ross's marker was to keep track of secular variations—long-term trends excluding periodic oscillations—in the relative level of land and sea. As with the markers Humboldt had proposed for the shores of the Caspian Sea, the one in Port Arthur was intended not for continuous observation but as a baseline. Having received Humboldt's suggestion only on his way back from Antarctica, Ross had not been able to put a similar marker on the rocks of the subantarctic Kerguelen Islands, which he had visited on his outward voyage. The lack of reliable data series like Lempriere's had also prevented Ross from installing a marker on the opposite coast of Tasmania. While bemoaning this, he expressed the hope "that some individual with like zeal for science . . . and with time at his disposal, may yet accomplish this desideratum." In December 1842 Ross set a further sea-level marker, deduced from five months of half-hourly observations, by Port Louis in East Falkland's Berkeley Sound.[5]

Central to Humboldt's vision was the desire to measure change over time on a global scale, an aim contiguous with that of Swedish scholars such as Anders Celsius who had engraved sea-level markers on rocks along their country's Baltic coast a century earlier. As Humboldt wrote of the markers he'd asked Ross to set:

> If similar measures had been taken in Cook's and Bougainville's earliest voyages, we should now be in possession of the necessary data for determining whether secular variation in the relative level of land and sea is a general or merely a local phenomenon, and whether any law is discoverable in the direction of the points which rise or sink simultaneously.[6]

Yet, the actual development of a global network of tide gauges took place in a context radically different from Humboldt's vision of a coordinated effort intended to create a reference framework for future comparative research. It came together, in fact, through an unsystematic, almost organic accretion that favored certain world regions over others. The idea

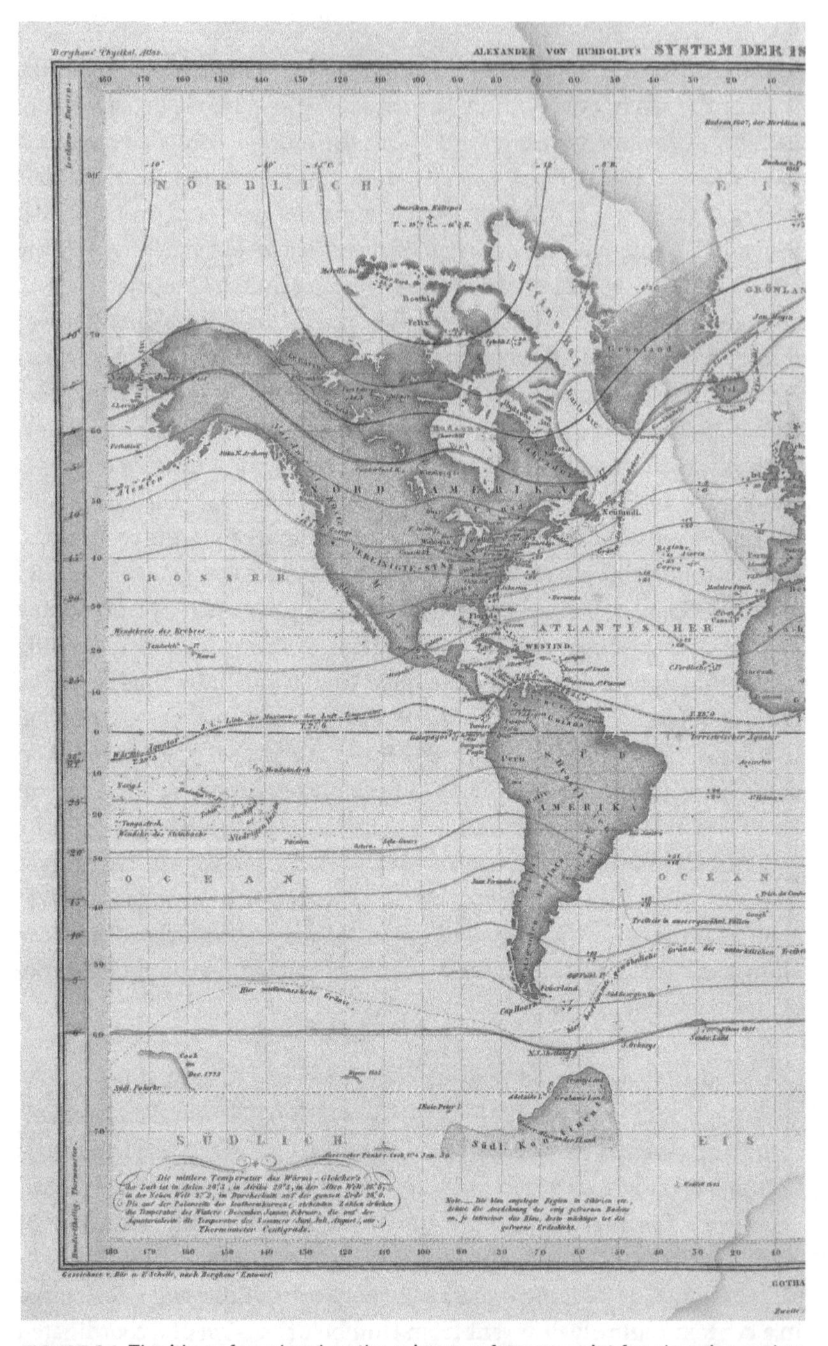

FIGURE 2.1. The idea of sea level as the primary reference point for elevation makes the world's coasts a kind of isometric line, one connecting the points at elevation zero. Alexander von Humboldt's insights into the use of isolines as tools of scientific inquiry prefigured later developments and was soon taken up and popularized by cartographers. *Source:* Heinrich Berghaus, "Alexander von Humboldt's System der Isotherm-Kurven in Merkator's Projection."

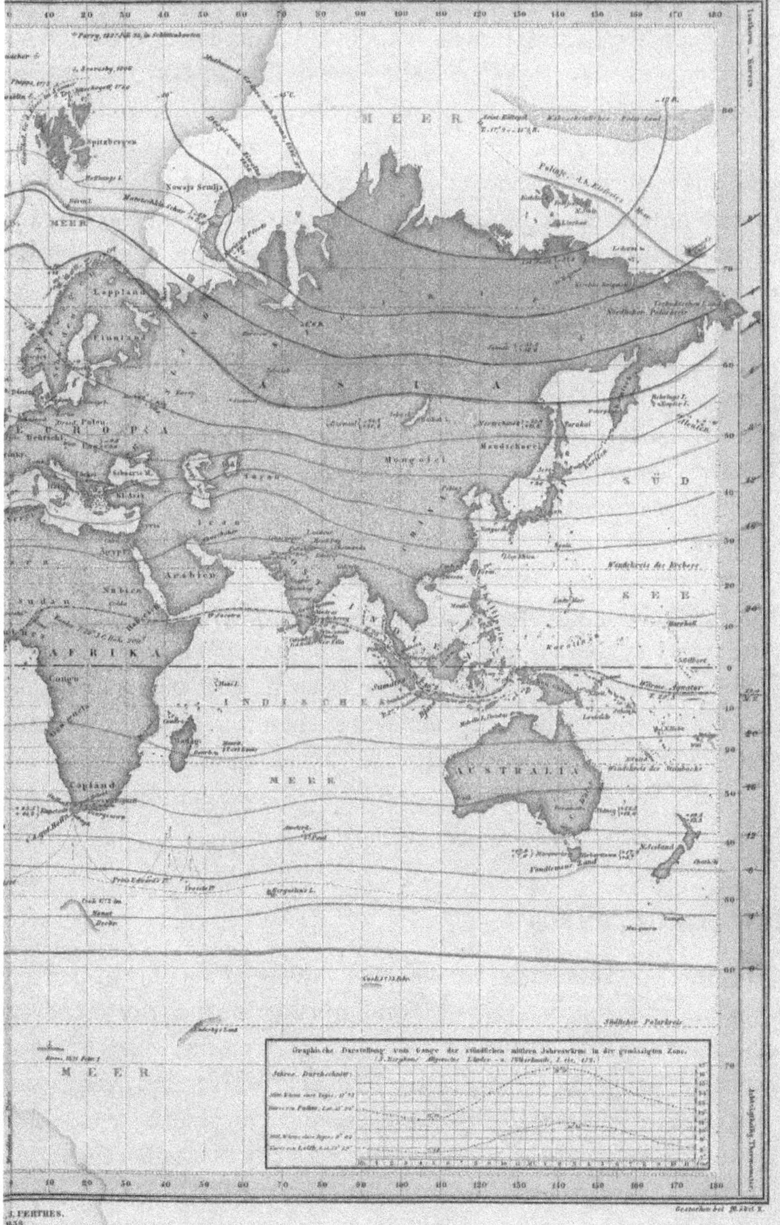

...OTHERM-KURVEN, IN MERKATOR'S PROJECTION.
Iᵗᵉ Abtheilung Meteorologie Nº 1.
Spitzbergen
Nowaja Semlja
MEER
Lappland
Finnland
EUROPA
ASIA
Mongolei
Mandschurei
Iran
Arabien
Aegypten
Nubien
Sudan
AFRIKA
Congo
Capland
INDISCHES
MEER
AUSTRALIA
SÜD
SEE
Wendekreis des Krebses
Wendekreis des Steinbocks
Nördlicher Polarkreis
Südlicher Polarkreis
Wärme Aequator
N. Seeland
Vandiemens Land
Madagascar
Sumatra
MEER
Graphische Darstellung vom Gange der stündlichen mittleren Jahreswärme in der gemässigten Zone.
Jahres Durchschnitt:
J. PERTHES.
Gestochen bei ...

of looking at the level of the sea historically would arise again and be translated into practical measures about a century later. By then, however, comparisons across time were necessarily based on observations whose distribution was determined more by practical and political reasons of state administration and colonial expansion than by scientific ones. The history of the assessment of sea level is part of a broader development toward the reform and unification of reference points and metrics that characterized the nineteenth century. This story is related in more detail in chapter 3.

It is also, as noted earlier, a story of infrastructural development. The needs of the infrastructural state, the spread of European science and technology through empire and colonialism, and the increasing interconnectedness of the world all played parts in the push to define usable reference points for elevations. The industrial revolutions spawned mobility and transport needs, which in turn drove an exponential growth of public works: canal networks were improved, railroads developed, and ports expanded. Public works were also quintessential in the making of colonial landscapes and new commercial routes, such as those cutting through the isthmuses of Suez and Panama. While most of these infrastructural developments were privately funded, they became possible thanks only to significant state investments in improved land surveys. These state-supported preparatory works fostered the impulse to define baselines and standardize methods for ascertaining elevations on land.[7]

DIFFERING LEVELS

Alexander von Humboldt was personally invested not only in creating a global network of markers that might contribute to the long-term analysis of sea-level changes. He was also interested in what he termed "comparative hydraulic hypsometry." *Hypsometry* is derived from the ancient Greek words for height and measurement and means as much. By hydraulic hypsometry Humboldt meant the study of whether the waters of the world's seas are part of the same surface and thus at the same level. Pursuing his usual belief that physical geography could only progress by regrouping existing data on a global scale to reach more general insights, Humboldt also recapitulated existing comparisons of interoceanic levels: across Panama and Suez, and between the Mediterranean Sea and both the Atlantic Ocean and the Adriatic Sea.[8]

The idea that there could be permanent differences in elevation among the world's seas—that they might not be part of the same surface level—had been repeatedly criticized by influential scholars. As early as 1752 barometric readings made by the Spanish scientific naval officers Jorge Juan and Antonio de Ulloa had suggested that the levels of the Pacific and Atlantic oceans were almost identical. In the 1780s the idea of disparate levels was criticized by John Walker, Regius Professor of Natural History at the University of Edinburgh. Yet the concept remained widespread, common knowledge among mariners.

In 1837, the *Nautical Magazine and Naval Chronicle*, the most important British journal offering practical knowledge to seafarers, founded earlier that decade by naval officer Alexander B. Becher, dedicated an article to the issue. Signed pseudonymously by Argonaut—probably Becher himself in his role as editor—the article sought, on the one hand, to provide theoretical backing for the selection of lowest low tide as representative of the sea's natural state and as a surface level, a view still preferred by seafarers and many surveyors at the time. The main argument was that, while water could be elevated by lunisolar attraction to cause the tides, no force could depress water below its natural state. Consequently this state had to be lowest low tide, the level "of the utmost possible depression of the water occasioned by the cessation of external influential causes." This was, however, a purely theoretical construct, not supported by geodetic or hydrographic measurements and offered as mere opinion for the consideration of the "master" who was working on the subject of tides—clearly a reference to Whewell. On the other hand, Argonaut took care to affirm that the physical properties of fluids precluded the possibility that parts of the world's seas could be "*permanently* higher or lower than other parts." He readily admitted that the level could appear higher in a minor inlet due to the influence of the winds. Such a phenomenon would, however, never produce a great difference in level and would cease as soon as the winds dropped. In larger gulfs this effect would be canceled out by marine currents. The perceived difference in sea level on the two sides of an isthmus, mentioned in multiple geographical works, were due, according to Argonaut, to local peculiarities. Citing the example of the isthmus separating the Bay of Fundy from the Gulf of St. Lawrence, he attributed the recorded height difference across the isthmus to the visual impact of the bay's extreme high tide and a lack of comparative studies.[9]

Humboldt had explained, almost thirty years earlier, how a similar impression—that the Pacific Ocean lies higher than the Atlantic—gained credit in the first place:

> This opinion is based on mere appearance. After having struggled for several days against the current of the Rio Chagre, one believes to have ascended much more than what he descends from the hills near Cruces to Panama. In fact, nothing is more misleading than the judgment made of the difference in level on a prolonged and consequently very gentle slope.[10]

Subjective comparisons of the effort expended in ascent and descent were thus deceptive. And barometric assessments made on the two coasts provided varying results, some indicating no difference in height, others that the elevations might differ by a few meters. The technical limitations of early attempts, such as those undertaken by Juan and de Ulloa around 1750, and the cursory nature of others, such as those made by Humboldt himself during his famous travels through South America, did not dispel conjectures that the Atlantic and Pacific oceans might be at significantly different heights. To resolve the question, noted Humboldt, would require a series of measurements made over the course of a year with high-precision, dedicated instruments and an awareness of such variables as the specifics of the tidal range on the two coasts.[11]

Another solution would be to complete a trigonometric survey of the isthmus. Exactly this was finally ordered around 1828 by the hero of Latin American independence, Simón Bolívar, then president of the Republic of Colombia, the state that he had founded in 1819 and that included present-day Panama. One of the surveyors was the same John Augustus Lloyd who would soon be instrumental in installing the first automated tide gauge at the Royal Docks in Sheerness. Aware of the confusion as to the exact definition of the "mean level" of the ocean, Lloyd adopted a fluid method. Rather than rest his assessment on one specific point or average—the method that became most common following Whewell's proposal that mean tide level be adopted as a proxy for the level of the sea at rest—Lloyd compared the levels of the two oceans at each step of their tidal progress:

> In every twelve hours therefore, and commencing with high tides, the level of the Pacific is first several feet higher than that of the Atlantic; it becomes then of the same height, and at low tide is several feet lower: again, as the tide rises the two seas are of one height, and finally at high tide the Pacific is again the same number of feet above the Atlantic as at first.[12]

While Lloyd imagined a point at which the two oceans were at the same level, he actually estimated a discrepancy of about 3½ feet (1.1 meters) at mean tide. Nonetheless, his survey finally dispelled the long-held conviction that the Pacific Ocean lies thousands of feet higher than the Atlantic. The French scholar François Arago observed that a difference of just over a meter, measured over a line of 132 kilometers and almost a thousand observation points in a mountainous territory, could be deemed to lie within the margin of error, "without insulting the merit" of Lloyd and his colleague, the Swedish captain Maurice Falmarc. Arago concluded that the difference in elevation, if there is one, is, as Humboldt later put it, "so small as to be inappreciable."[13]

Not all such questions, central as they were to the development of transportation networks, found such swift clarification. The debate over the relative elevation of the Atlantic and the Mediterranean went on for decades (as discussed in chapter 3). Similarly, attempts at comparing the elevations of the Mediterranean and the Red Sea, in advance of the planned excavation of a canal through the Suez isthmus, continued throughout the first half of the nineteenth century.

SURVEYING SUEZ

At the turn of the nineteenth century, during the French invasion of Egypt, Napoleon's chief surveyor, Jacques-Marie Le Père, claimed there was a difference of about ten meters between the water levels on the two coasts of the Suez isthmus. The idea that the Red Sea lies higher than the Mediterranean dates back to antiquity and is cited by, among others, Aristotle. Most of the related debate seems based more on supposition than observational data. Multiple historical accounts, ethnographic evidence, and archaeological excavations hint at the presence, until the early middle ages, of a canal connecting the Bitter Lakes north of Suez and the Nile, the so-called Canal of the Pharaohs. Its mere existence would seem to disprove a difference in level, but many ancient accounts claim the canal was never finished because of the drop. Actual sea-level gaugings on either side of the isthmus were started only in the eighteenth century. The first assessments of tidal movements in the area were apparently made in 1762, on behalf of the Danish crown, by the German explorer Carsten Niebuhr, who like Le Père at the end of the century, focused on the amplitude of the tide and the times of its extremes rather than on mean tide level. The first data series to record the latter, if only cursorily, seem to have been produced by the French state topographer Paul-Adrien Bourdalouë in 1847.[14]

One aim of Napoleon's Egyptian campaign was to create the conditions to cut a canal, a more direct connection between Europe and the Indian Ocean that would foster French commercial interests. Clarifying whether the two seas were at the same level was crucial, as the scientists involved well knew. Jérôme Lalande, for instance, while skeptical of the long-held belief in different elevations, reminded the expedition's astronomers to look into it. At the end of 1798 Napoleon himself, accompanied by a scientific delegation, visited the isthmus and the vestiges of the Canal of the Pharaohs. Once back in Cairo, he ordered Le Père to start the land survey as soon as possible. Le Père's initial reference point, recorded on January 25, 1799, was high water in the port of Suez. From there his officers proceeded with their spirit levels north toward the Bitter Lakes, following the bed of the ancient canal. The work was interrupted, however, after just ten days, having covered about a quarter of the overall distance, due to lack of drinking water. Amid the ongoing military campaign, the survey could not to be resumed until September but then proceeded swiftly: with an average of about 500 meters between each station, the surveyors covered up to 12 kilometers per day. Overall it took them only twenty-five working days to survey the whole isthmus. The difficulties of working in a desert without much prior knowledge of the territory during a war of occupation did not leave time for thorough verification. The work coordinated by Le Père was thus subject to intrinsic limitations, including the fact that it referred to high water and that the elevation of the sea on each side of the isthmus was measured only once, not averaged over a longer period of time.[15]

In his initial account of the survey, Le Père attributed the apparent significant difference in altitude between the two seas to the impact of easterly winds over the Indian Ocean. In 1809, when his text was included in the state-sponsored *Description de l'Egypte*, Le Père attempted no explanation of the result, preferring to leave that task to others. While almost no one asserted publicly that Le Pére had made mistakes in the leveling operations, his results were widely criticized behind closed doors. The committee in charge of publications turned down attempts at theoretical justification of his results. In particular, Pierre Simon Laplace and the mathematician Joseph Fourier were skeptical.

Nonetheless, Le Père's work gained a following among engineers and others concerned with practical aspects of the potential canal. It was taken at face value, too, by scientists such as François Arago. It appeared almost unreasonable to doubt the precision of the leveling work done by

FIGURE 2.2. Surveying in a desert during an armed conflict limited surveyors' ability to perform their work reliably and accurately. Nonetheless, the reputation of the French Army underwrote the results of Jacques-Marie Le Père's survey of elevations across the Suez isthmus, which for decades remained the golden standard. A scene depicting the leveling works north of Suez highlights the isolation of the setting. *Source:* Le Père, "Plan et nivellement des source dites de Moyse," plate 10.

officers of the French army, especially when it agreed with the received wisdom of ancient philosophers. The status of Le Pére and his acolytes, and the context within which their work was accomplished, enhanced the perceived reliability of the results. There was no lack of projects for canals. Some proposed using entire lakes as locks to overcome the difference in elevation. Others posited that a sloping canal would be beneficial, as the steady, permanent current would avert the risk of silting. For almost half a century, however, fear that opening a channel to the Red Sea might submerge land and ruin the water of the Nile stalled all attempts at building a canal.[16]

Finally, in 1847, a new survey was undertaken by Bourdalouë, on behalf of a multinational consortium and with the support of Egypt's ruler, viceroy Muhammad Ali. The support of the viceroy ensured that the work would not be subject to disturbances like those Le Pére had faced. Huge investments were made and great care taken to adopt methods that would maximize precision. Bourdalouë deployed two teams to survey the region in parallel, and within each team had elevations read separately by two people. If the results within a team diverged by more than 2 millimeters the work had to be redone. Every couple of kilometers the two teams would meet, and if their measurements of a common landmark differed by more than 12 millimeters, the whole section would be redone. The whole survey was done twice—certain sections three or four times—until all levels along the route were double-checked. One of Bourdaloué's collaborators then made a second, rough survey to confirm the results. Bourdaloué's outcomes refuted Le Père's. At low tide the two seas appeared to be at almost exactly the same level. At mean tide, a difference of about 80 centimeters could be established, which was less than the tidal range of the Mediterranean Sea. Once again, however, Bourdaloué determined the reference point, as had Le Père, on a single day. And he too adopted a tidal extreme as the starting point of his survey: on the Mediterranean coast the selected datum was low tide at Tell el-Farama, the ancient Pelusium, on December 8, 1847; on the Red Sea it was low tide at Suez on November 25. For context, Bourdaloué also provided the high-water marks, as well as some information on the range of tidal extremes on the Red Sea, gathered from both British navy officers and the harbor master in Suez. Full series of tidal fluctuations, which would allow ascertainment of mean sea level, were recorded only later, once preparations for excavation were started.[17]

Bourdaloué's results ignited a lively debate. Even if he had made all possible efforts to ensure his survey's precision, not everybody was ready to accept that Le Père had erred, or to refute the theoretical framework his decades-old results had shaped. Even the French engineer Maurice Linant de Bellefonds, director of public works for Upper Egypt, who had taken part in Bourdaloué's survey, had difficulty accepting that Le Pére had made such a significant error. To quell his own doubts, in 1853 Bellefonds decided to resurvey the area. Once his effort confirmed Bourdaloué's, acceptance spread and concern over a drop between the two sides of the isthmus diminished. A year later Ferdinand de Lesseps was charged by the Egyptian authorities to start work on the construction of a sea-level canal.

Truth be told, other surveys showing that the two seas were at the same level had already been performed in the intervening years. For a variety of reasons, both political and technical, none of them had gained enough credence or diffusion to successfully challenge Le Père's results. Neither work done by European surveyors on behalf of the Egyptian authorities nor that done using barometers had been accepted as on par with the geodetic leveling performed by the officers of Napoleon's army. Bourdaloué, however, was himself a French civil servant. And he had resurveyed the area by geodetic leveling. Moreover, he had taken great care to incarnate a new way to certify scientific expertise liberated from deference to ancient philosophers.[18]

THE MAKING OF REFERENCE POINTS

While the relative elevation of seas continued to be debated, national standard gauge points were defined in a relatively short time throughout Europe once mean sea level was formalized as a reliable proxy of the geometrical shape of the planet. Modern, self-registering tide gauges were not yet universal; probably because of their cost, up to a third of the gauging stations along Europe's coast still used simple scaled gauges in 1890. These included the historic gauge in Amsterdam, which, as will be discussed in more detail in chapter 3, had recently become the basis of the German ordnance datum. All other stations charged with determining national datums, such as Marseille, Genoa, and Trieste, had by then adopted some form of *mareograph*, as automated tide gauges were known in continental Europe.

Politics played as much a role as scientific insight and reliable observations in establishing how national datums were first selected. Indeed, politics often proved more important than the accuracy of measurement. Before 1860 a number of arbitrary reference levels had been used for different regions of France. In Paris, heights were ascertained in relation to a horizontal plane passing about 75 meters below the zero of the scale at the Tournelle bridge, connecting the east bank of the Seine with the Île Saint-Louis; for the Marne, a plane was specified almost 115 meters below the same point. Other departments adopted unrelated standards: for the Loire, the level of the Atlantic Ocean at Saint-Nazaire, near Nantes; for the Cher and the Allier north of the Massif Central, a landmark on a pillar of the Bourges cathedral; for the Rhône, the level of water at low tide in the port of Marseille. This lack of uniformity led to confusion, especially when networks overlapped. Even markers located side by side offered no apparent basis for comparison. In this instance, as in many others, improvements to the national transportation networks compelled the adoption of uniform geodetic baselines and standards. In France the

FIGURE 2.3. Prior to national standardization in the 1860s, authorities in various French regions marked multiple locations as elevation reference, creating a mishmash of overlapping geodetic frameworks. The one in Paris was defined as the plane passing about 75 meters below a marker engraved near the Tournelle bridge, which was defined as the level of the Seine during the drought of 1719. Image: Johan Jongkind, *Notre-Dame vue du quai de la Tournelle* (1852), Dutuit collection, Musée des Beaux-Arts de la ville de Paris.

lack of a single national benchmark for elevations became particularly pressing from the 1850s onward, with plans to extend existing railways and create an integrated rail network.[19]

One of the main aims of the new French national geodetic survey—directed between 1857 and 1863 by the same Paul-Adrien Bourdalouë who in 1847 had resurveyed the Suez isthmus—was to select a standard national vertical datum. Initially, the reference plane was the mean level of the Atlantic Ocean in Saint-Nazaire, on the estuary of the Loire. But this choice soon came into question: when measurements made at different ports on the Atlantic Ocean and the English Channel were connected through the leveling network, surveyors found that mean levels varied significantly. It seemed difficult to justify selecting the level at one location on the Atlantic as the national datum. The lesser intensity of tides in the Mediterranean and the greater consequent stability of mean sea levels along that coast led surveyors and government officials to favor the latter as the site of a new national vertical reference point. Furthermore, choosing the Mediterranean, the mean level of which seemed to be up to a meter below that of the Atlantic Ocean, ensured that all altitudes could be expressed with positive numbers. On January 13, 1860, the minister of public works, Eugène Rouher, adopted as the new national datum the plane passing 40 centimeters above the zero mark of the tide gauge at Fort Saint-Jean in Marseille, which he deemed representative of the mean level of the Mediterranean Sea. Little is documented about the process by which the marker was determined. One gathers that, once more, it was an estimate rather than a measured level.[20]

In Great Britain, too, the datum for the country's first geodetic leveling was at first estimated rather than measured. There, however, authorities shifted swiftly toward a thoroughly measured mean sea level. In 1840, when the leveling project was started, the reference plane was arbitrarily defined as one passing 100 feet below a benchmark on St. John's Church in Liverpool. But the work done by Airy in 1842 to select a datum for the Irish survey soon spurred change, and in 1844 a new marker was set on Victoria Dock in Liverpool, on the mouth of the Mersey. In March of that year tidal extremes were observed for ten consecutive days on a tide pole: every five minutes, for an hour around high and low water, officers recorded the water level. The new marker stood 43.14 feet above the previous reference plane, rounded down to 43 to make it easier to convert heights.[21]

The debate about the accuracy and reliability of the national vertical datum did not end there: in 1859, toward the end of its twenty-year leveling effort, the Ordnance Survey coordinated the recording of sea level at thirty-two locations around Great Britain. Despite the spread of automated tide gauges, most of the data were still collected as visual observations following Airy's instructions: for a fortnight officers checked a tide pole every ten minutes from slightly before one daytime high (or low) tide to slightly after the next. Every weekday, that is, they made about seventy-five observations in each location. Data for some locations, however, were later deemed of insufficient quality to include when calculating a national mean sea level to comparatively evaluate the Liverpool benchmark. These included data collected in Portsmouth and Plymouth, probably using the self-registering gauges installed by the Admiralty. The only automated tide gauge whose measurements seem to have made it into the computation was one that had begun operating at George's Pier in Liverpool in early 1854. The observations made there yielded a mean tide 8/10 of an inch (about 2 centimeters) above the reference point established nearby fifteen years earlier. This exercise—excluding the stations on the Thames and those whose data were considered problematic—showed that the level of the sea around the coasts of Great Britain was about 7½ inches above the Ordnance Survey datum. This result highlighted the volatility and arbitrariness of any national datum, however selected. It thereby consolidated the idea that a reference benchmark is primarily a convention, valid as long as it is deemed useful by the state administration and its surveyors. Datums are technical means, the products of scientific and political processes. Even when they refer to apparently natural markers, they remain subject to constant revision.[22]

MEANS OF COLONIAL SURVEYING

Expanding transportation networks were central factors when selecting reference points not just in the metropole but in European colonial territories and protectorates around the world. These disparate settings brought additional difficulties: the technical and physical challenges of mapping unknown territories, limited financial resources, competition among colonial agencies. Since 1802 the datum used throughout the territories controlled by the East India Company had been low water, as determined at various local tide gauges. In the 1820s and 1830s the officers in charge of improving the canal network in Bengal still referred to assessments of low-water level made in 1824 by the master shipbuilder James Kyd in Calcutta (now Kolkata).[23]

In early nineteenth-century India, elevation was, in most instances, defined barometrically. An officer would observe the height of mercury on a barometer and compare it with a reference level. No complex measurement infrastructure was needed, only handheld instruments and conversion tables. The push for surveys in the Himalayas stimulated adoption of this relatively rapid method, though it was widely considered exceedingly imprecise. And barometers, while simple, were fragile, difficult to carry far afield. Boiling-point thermometers were cheaper and less delicate, but the readings they produced were even less accurate; plus, they required that a fire be set up, which in extreme environments with little wood sometimes made them unusable. Nonetheless, mountain barometers and boiling-point thermometers had become increasingly popular by the 1820s, often favored over trigonometric height assessments, especially in the absence of a continuous surveying network.

Typically, the instruments were used to ascertain the elevation of a base station, from which further trigonometric surveys could be undertaken. The assumed reference level was the altitude as indicated by the barometer held at the Surveyor General's House in Calcutta, which in turn was related to Kyd's 1824 determination of low-water level. Whether the difference in altitude between the Surveyor General's House and sea level was accounted for in any way in previous surveys is unclear.[24]

Until the end of George Everest's term as surveyor general in 1843, tidal observations played only a minimal role in the overall work program. In 1834, the court of directors of the East India Company wrote that it had no intention "to incur any considerable expense in pursuing the object of science." Tide registers would be kept only when it could be done cheaply and without inconvenience. The task, the directors felt, was best left to the voluntary efforts of "scientific men"—a stance that continued a long tradition, rooted in early modern science, favoring the work of natural philosophers over that of practitioners. Even at a time when science was increasingly being professionalized, the institutional take seems to have been that gentlemen scientists motivated by passion would produce, if not more accurate, at least perfectly adequate data—and without the cost of paid personnel. In reality, the role of technicians and salaried observers was as crucial in this as in other fields of science—recall the many contributors to Whewell's work on tides, discussed in chapter 1—though it was underrecognized by both scientists and government officers at the time.[25]

Obviously, tidal observations were not completely unknown, or even uncommon, especially as means to plot and understand shifting coasts and riverscapes. They had been carried out continuously in Calcutta on the river Hooghly since 1806, while onetime surveys were made in Madras (now Chennai) and Chittagong (now Chattogram in Bangladesh) in, respectively, 1821 and 1837. In March 1835 Everest first called for connecting the subcontinent's geodetic network to sea level. More observations were carried out, usually for rather short periods of time, along the coast of India by navy officers, most motivated by purposes unrelated to the survey or by the curiosity of individual assessors or men of science.[26]

British colonial officials were, nonetheless, eager to try out the recent advancements in sea-level gauging. As early as 1839 representatives of the Madras Observatory visited Bristol to see the self-registering tide gauges that were being built for the East India Company. In 1842, one of these gauges was set up at the Colaba Observatory in Bombay (now Mumbai).

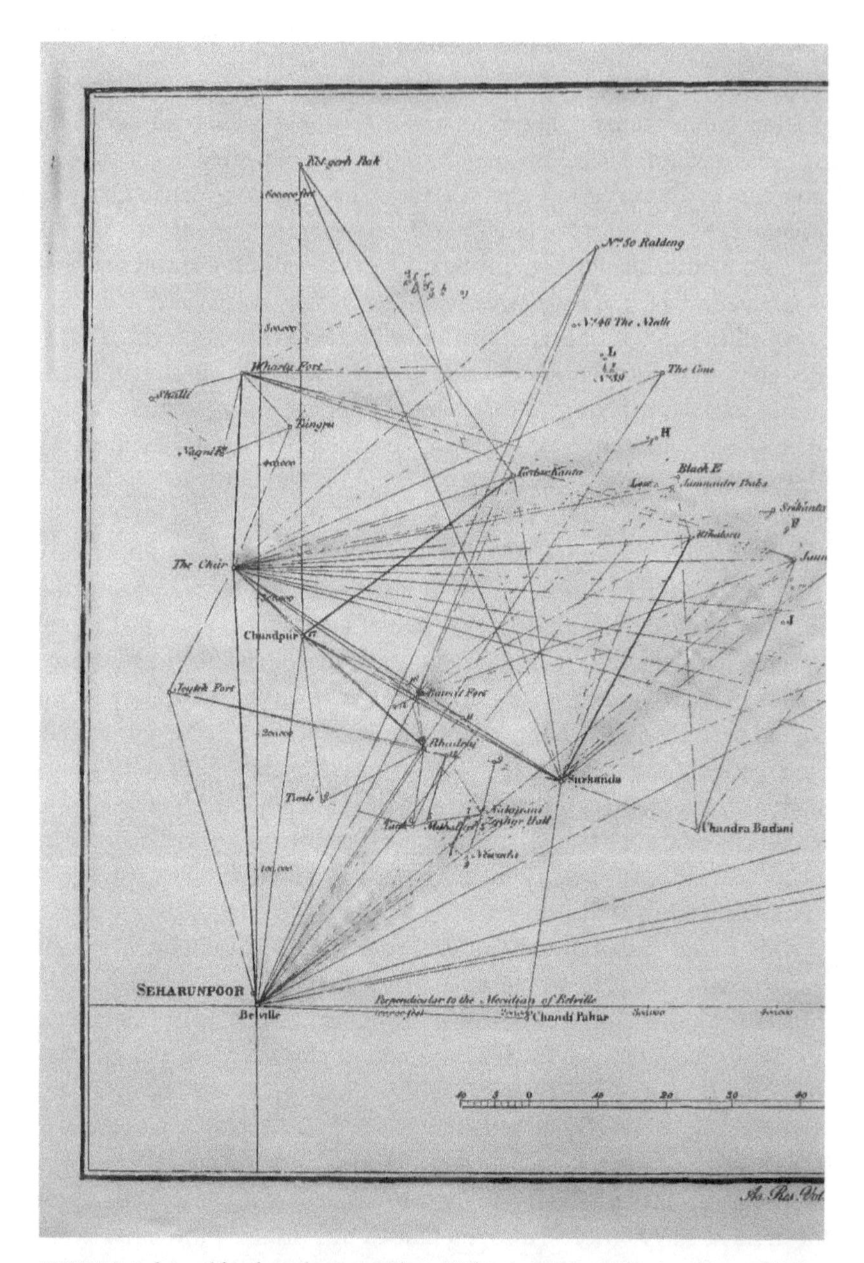

FIGURE 2.4. A combination of barometric and trigonometric methods was used in the earliest stages of the British survey of the Himalayas. While the elevation of the Belville station in Saharanpur was determined barometrically, from there peaks were measured through triangulation. The hybrid process arose from a combination of causes: on the one hand, no continuous surveying network as yet extended to Belville; on the other, barometers were too fragile to be safely brought up to the peaks, and many peaks had not yet been reached. *Source:* Hodgson and Herbert, "Account of Trigonometrical and Astronomical Operations," plate III.

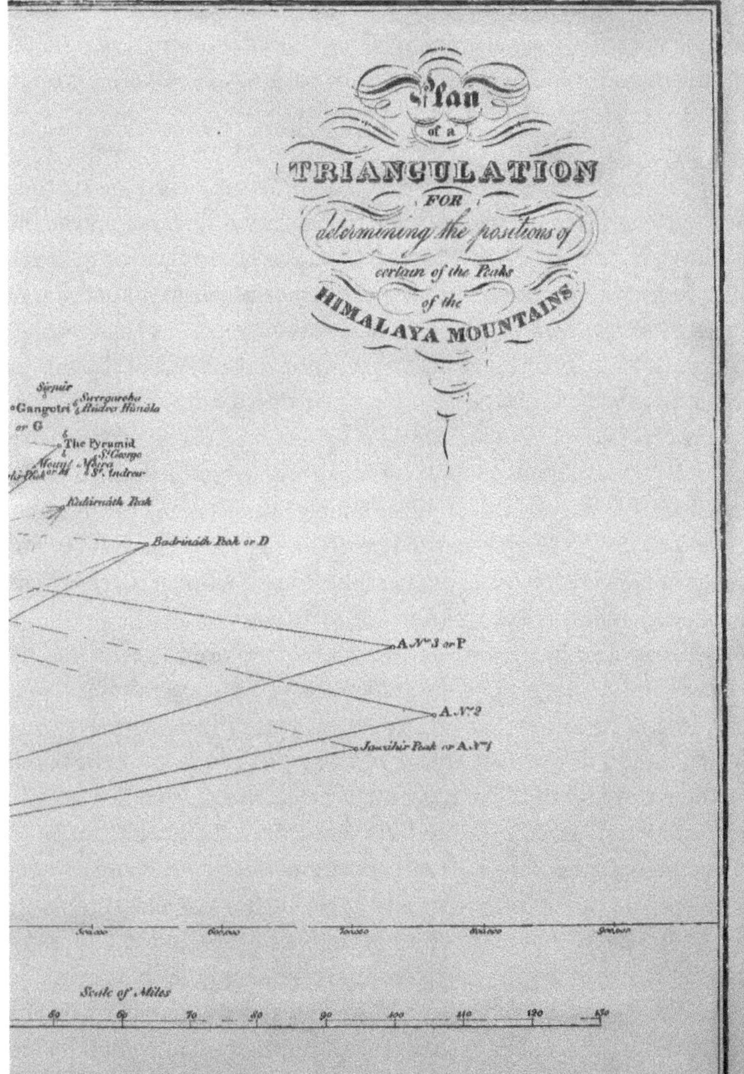

Plan of a TRIANGULATION FOR determining the positions of certain of the Peaks of the HIMALAYA MOUNTAINS
Gangotri or G
The Pyramid
St. George
St. Andrew
Kedárnáth Peak
Badrináth Peak or D
A. No. 3 or P
A. No. 2
Jawáhir Peak or A. No. 1
Scale of Miles

Founded by the Company in 1826 as an astronomical observatory, it ended up serving over the years the practical needs of surveying and navigation in the region. Two more self-registering tide gauges were set up in Karachi (in current Pakistan) and Porbandar around 1850. In the years that followed—with increased attention to accuracy in vertical measurements spurred by Whewell's redefinition of mean sea level—the attitude of the survey's officers toward the need to determine a reliable and stable reference plane changed radically. In 1845, the new surveyor general, Andrew Scott Waugh, began making explicit requests to his officers to verify heights "by reference to the direct level of the sea." By then—about ten years after Whewell discussed in print the mean level of the tide as the standard to be used in ascertaining the vertical datum—sea level was, as reported by Waugh in 1853, "universally assumed to correspond with mean water." The methods adopted by the survey in establishing the level of the sea were, however, still of questionable precision: in one instance, only five data points were recorded between high and low tide; in another, agreement within 3 inches over four tides of manual observations of a pole stuck in low water was deemed satisfactory.[27]

Waugh was also the first surveyor general to express "great anxiety concerning the . . . heights," especially in regard to ascertaining elevations in Northern Hindustan. In support of this difficult undertaking, Waugh planned a new survey that would connect the Bay of Bengal and the Arabian Sea and thus the two coasts' reference planes. The importance of the series, asserted Waugh, required, first, the establishment of an independent datum "beyond cavil or dispute" on the Arabian Sea. Surveyors were sent out to determine a suitable site for tide measurements on the Gulf of Khambhat (then Cambay) in Gujarat. Their attempts were unsuccessful, because the multiple estuaries opening in this bay made it hard to distinguish between fresh and salt water and to make sure the gauges were not affected by the rivers. In the end the connection had to be set up farther west, in Karachi, where the measurements performed by the survey were connected to those made at Manora Point lighthouse by naval authorities; later, a self-registering gauge was installed under the supervision of the Bengal Engineers lieutenant James Tennant.[28]

When David Nasmyth, an officer in the Bombay Sappers and Miners assigned to the Great Trigonometrical Survey, finally found a location on the Gulf of Khambhat that seemed fit to be used to ascertain a datum, the naval draftsman at the Colaba Observatory contested the choice, claiming that Nasmyth had not considered the navy's needs for the precise calculation of the time of tides. Waugh supported the objection and stated that the criteria adopted by land surveyors in choosing where to locate the tide gauge were too limited and biased. Their exclusive focus on ascertaining mean sea level led them, said Waugh, to attend only to convenience of access, lack of excessive waves, and the overall stability of the gauge's zero, resulting in choices that proved unsuitable for the "general purpose of science." The interests of the navy and those of the survey in regard to siting gauges differed substantially, but Waugh insisted that the selected locations serve the needs of both.[29]

In 1852 the colonial government asked Waugh if it would be possible to mark the height above sea level of "all *obligatory points* throughout India, having reference to the construction of works, draining, irrigation, and so on, such as the highest points or necks of valleys and the low passes of low ranges." Waugh pointed out how expensive, labor-intensive, and—all things considered—useless such an endeavor would be. Precise local plans would always be necessary for special purposes, like irrigation projects, but such detail was uneconomical as part of the general maps produced by the surveying effort. The government did not press the request, but in 1855 the Punjab canal department pleaded with the survey's officers for a series of leveled benchmarks (a request that would not be fulfilled until 1903).[30] When discussing reference planes at varying geographical scales, from the local to the international, different exigencies needed to be satisfied. A multiplicity of datums is, in fact, frequently necessary and convenient to cover the needs of a variety of actors, such as surveyors, entrepreneurs, farmers, and the military. As shown above, national datums were first defined to facilitate the expansion of transportation networks to supraregional scales and to strengthen control over colonial territories by providing a common frame of reference. Many practitioners deemed them superfluous nonetheless, because often what mattered in specific local contexts were relative rather than absolute elevations. The choice to impose a certain reference system is essentially political.

Once mean sea level became widely accepted as the main benchmark for altitudes, multiple efforts were made to produce reliable series of observations and select apt locations for tide gauges. The logistical needs of rapidly developing national economies and the surveying necessities of global colonial endeavors commanded the spread of gauging stations and the propagation of automation—an uncoordinated diffusion that, favoring some areas of the globe over others, created the preconditions for what would be, in the long run, a skewed view of how sea level varies.

Alexander von Humboldt had envisioned a global network of markers recording a baseline for the analysis of sea-level changes. This could have been the beginning of a measurement infrastructure based primarily on scientific criteria. Efforts in this direction were, however, few and disconnected. In reality, multiple specific interests determined where and when new stations were installed and how the infrastructure developed.

The material bases for standardizing the assessment of elevations were both directed and limited by national politics, colonial efforts, and international competition. French authorities kept the estimated sea level they originally chose as their national datum, while the British changed theirs to reflect technological advancements. Both were political decisions; neither favored a more "natural" baseline or evinced a more scientific method. The resistance to change and innovation in the Suez case and the continual interagency conflict in British India were likewise products of particular political conditions. Similarly, as will be discussed in detail in the next chapter, attempts to attain an *international* reference plane had to compete with existing national ones by demonstrating some added value. Surveyors were employees of the state, and for most of them propagandizing the virtues and advantages of national benchmarks amounted to a natural defense of their corporate interests.

THREE

STANDARDS OF HEIGHT

In a booklet published in 1931 titled *Marseille, ou la mer qui monte* (Marseille, or the rising sea), Auguste Bouchayer, an entrepreneur and engineer from Grenoble, reports on his hunt for the technical means by which the mean level of the Mediterranean at the port of Marseille is determined. After a few bureaucratic dead-ends and a tram ride through town, Bouchayer gets to the automated tide gauge, or "integrator," that was set up in 1884 in the city's harbor—and realizes in horror that the data it's collected over almost fifty years has been corrupted. The wife of the gauge's warden, looking after the machinery while her husband is out fishing, reveals to Bouchayer and his companion—Louis Le Doyen, an amateur archaeologist working at the city's roads and water administration—that when the sea is too strong, they just close the lower basin. Then they can sleep without fretting that the float's cable might break. But by doing so they effectively impede the gauge's ability to record continuous measurements. Bouchayer expresses his dismay: "My dear Le Doyen, your integrator, marvel of mechanics, is rigged! Our bases are precarious!"[1]

Bouchayer's interest in sea-level change came unexpectedly. He worked as an engineer on the water intake dam for a hydroelectric power plant being built at the confluence of the Romanche river and the Drac, a tributary of the Isére, 200 kilometers from the Mediterranean Sea, at an elevation of about 250 meters. In this capacity, he had come across archeological and paleontological findings that got him thinking about the radical changes that had occurred over the millennia in both the flow of Alpine streams and the conformation of the Mediterranean coast. In particular, he got interested in the "tangible marks of the great periodic movements." Now obscure—its theories about a sea rising due to cosmic influences did not stand the test of time—Bouchayer's book was widely reviewed in the year following its publication by a number of French newspapers and magazines, testifying to some interest among the French reading classes in the question of how coastlines came to be.[2]

Precariousness is intrinsic to all measured datums, as they depend on the constant changes experienced by the sea. (Witness the changing British datum discussed in chapter 2.) This realization led German surveyors to prefer for the new German Reich—the state set up in 1871 as a consequence of the national unification process led by Prussia—a datum unrelated to any natural feature of the coast. While this approach allowed a degree of long-term certainty in map-making, it met its own limitations when the new country became a colonial power a little more than a decade later. Each of the Reich's colonies ended up adopting a different datum, reflecting the particular needs of local administrations. Each, however, was a measured benchmark referring to the level of the sea, assumed to be at the same level as the reference plane adopted in the homeland. The divergent approaches adopted in the German metropole and colonies exemplifies once more how sea level had become, by the end of the nineteenth century, the de facto global standard reference plane for elevations.

Nonetheless, scholars and surveyors persisted in attempts to more formally define a standard datum. In particular, a debate over how to determine a common European point zero of elevations occupied representatives of various countries through the second half of the century. This debate, driven by participants' attempts to establish their own countries' datums as the continental standard, ended by reasserting that mean sea level, ascertained at various locations along Europe's coasts, was the best approximation of a common reference point. While the discussion did not lead to a standardized continental geodetic framework, it reinforced the idea that datums are conventions, whose choice is linked more to political power than to the accuracy of measurement or the strength of scientific theories.

A DATUM FOR THE REICH

The series of data recorded by the Marseille integrator between 1885 and 1897, widely publicized in the press for its precision, was used to calculate a new reference point that replaced the marker at Fort Saint-Jean, set years earlier as the national vertical datum. By 1883 fifty-seven self-registering tide gauges had been installed along the coasts of Europe. During the same period, various countries adopted iterations of mean sea level as national vertical datums. Austria-Hungary selected the mean level of the Adriatic Sea in Trieste as measured over a year in 1872. By 1890 Italy and Spain had selected the mean level of the Mediterranean as ascertained, respectively, in Genoa and Alicante, while Portuguese authorities had chosen the mean level of the Atlantic Ocean recorded in Cascais, and Romanian surveyors had settled upon the mean level of the Black Sea in the port town of Constanța. Russia, considering the extent of its territory, adopted the average of the mean levels of the Gulf of Finland in Kronstadt, by St. Petersburg, and the Black Sea in Odessa. Throughout the 1880s and 1890s most of the stations chosen to determine nation-states' vertical datums were equipped with automated mareographs, warranting improved accuracy in their measurements.[3]

Mean sea level was thus widely accepted throughout Europe as the standard reference point for elevations. It did not, however, immediately become the sole reference point for elevations on land. In 1879, the Belgian civil authorities adopted as a vertical datum mean low water at spring tide in Ostend, so that surveyors would never have to record negative elevations even when working in liminal areas along the coast. The Netherlands continued to prefer the mean of high water in Amsterdam, recorded continuously since the seventeenth century, as its main reference point. In what was then still Prussia, a new precision leveling was begun in 1864. Different reference points were used in different parts of the expanding country. The eastern provinces referred to the gauge zero in Neufahrwasser (currently Nowy Port, a suburb of Gdańsk), a point set below the level of the lowest low tide, about three and a half meters below mean sea level. The western provinces referred mostly to the Amsterdam ordnance datum but occasionally to the mean level of water in a port on the North or Baltic Sea. Frequently the chosen port was Swinemünde (now Świnoujście in Poland), where reliable non-automated data had been collected regularly since at least 1826, making it the longest available series on the Baltic Sea. In 1835, on the basis of measurements taken thrice daily between 1826 and 1834, Prussian surveyors determined mean sea level in the port, a measure then used as a reference plane for some of Prussia's leveling operations. A self-registering tide gauge was finally installed in 1870, allowing mean sea level to be calculated from the continuous curve of water levels rather than individual measurements. For good measure, the station was connected by leveling to the Amsterdam ordnance datum and compared, in cooperation with survey officers in the duchies of Mecklenburg-Schwerin and Mecklenburg-Strelitz and the free city of Lübeck, with data from the stations in Travemünde, Wismar, Warnemünde, and Stralsund, all on the Baltic coast. This comparison led to the hypothesis that the level of the Baltic Sea slopes by about half a meter between Memel (now Klaipėda in Lithuania) in the east and Lübeck in the west and, consequently, to a wide mistrust among German geodesists for the usefulness of a single measured mean sea level as the vertical datum, especially given the perceived limits of existing instruments and the way they were interconnected.[4]

Driven by the desire to integrate improved geodetical data on the horizontal plane with a clear reference frame for the vertical dimension, the new German Reich instituted in 1875 a commission charged with defining a unified datum. Any single, local mean sea level was immediately ruled out, since the invariability of mean sea levels at points along the coast had never been proven. The commissioners deemed Whewell's assumption of uniformity to be "theoretically almost unthinkable."[5] Furthermore, the eccentricity of all coastal locations in respect to the leveling network argued against basing the new datum along a coast. They preferred instead a place located centrally in respect to the network and steeped in old alluvial ground, thus presumably less subject to geologic elevation or subsidence. Coincidentally—or maybe not—Berlin, the Reich's capital, seemed to fit the bill.

Deciding that the marker should be located inland did not, however, solve the issue of how to establish the datum. Even if the commission's members agreed that, in keeping with common use, some kind of mean sea level was preferable, they regarded the concept as impractically vague. Long-term measurements made at tide gauges along the coasts had provided such varied mean sea levels that no single equipotential surface could be reliably ascertained. Moreover, the science of the time did not allow one to ascertain the average level at high sea, and thus to clarify whether the sea was, at different locations, actually at the same level. An average altitude of the sea could be established only as an approximation or by convention. The majority of the commission agreed that the Amsterdam datum—already commonly used in Prussian surveys and broadly consistent with measured mean sea levels along the German coasts—was the best available approximation of the "average altitude of the sea" and, consequently, an ideal reference point for the leveling network.[6]

In 1879 the Prussian authorities adopted a new, unitary vertical datum, the *Normalnull*, defined as the equipotential surface passing 37 meters below a marker at the Berlin Observatory, the *Normal-Höhenpunkt*. This benchmark, a transposition through leveling of the Amsterdam ordnance datum, was an answer to the *Horizontfrage*, the question of what geodetic horizon should be used, an issue that had vexed the German geodetic community for decades. The persistence of different datums in the states that came to form first the Norddeutsches Bund and then the new Reich, and even within Prussia itself, had repeatedly spurred calls to create a nationwide reference plane, a *Reichshorizont*.[7]

That the new datum was a convention, unrelated to any "natural" feature of the planet, was clear to the members of the commission. That it actually differed from the Amsterdam datum, as more precise levelings would reveal, made no difference at all. The new Prussian datum was soon adopted by other states within the German Reich. Within a few decades following the definition of mean sea level as the best approximate marker of an equipotential surface level, all European countries had selected a unique vertical datum for their leveling efforts. But the process by which one was selected in Germany diverges radically from the approach adopted in France and a plurality of others: German surveyors stated outright that the national vertical datum would be a convention, rather than claiming that it was a measurable, "natural" reference point.[8]

Die Königliche Sternwarte zu Berlin.
von der Nordseite gesehen.

FIGURE 3.1. The Prussian *Normal-Höhenpunkt* was expressed materially as a marker embedded in the north wall of the Berlin Observatory, erected in 1835 in the Kreuzberg neighborhood. The marker was placed 37 meters above the country's new datum, defined in rapport with the Amsterdam datum, considered the most stable due to the length of the data series on which it was based. *Source: Der Normal-Höhenpunkt für das Königreich Preussen*, plate I.

SURVEYING AN EMPIRE

After decades of discussion among scholars and practitioners about how to reliably determine mean sea level and the possibility of using it as an international benchmark, the blunt statement that the Prussian *Normalnull* was a mere convention effectively decoupled the geodetical framework from measured sea level. When surveyors and state officials in other countries opted to keep mean sea level as their national vertical datum, it was equally clear that this was a political choice based mainly on issues of convenience.

Nonetheless, even as *Normalnull* was adopted by other states within the German Reich, it didn't become the new state's only vertical datum, as had been expected. And Germany's late nineteenth-century venture into colonial imperialism complicated the matter. From 1884 onward, German colonies—from South-West Africa (current Namibia) to the Kiaoutschou Bay concession in northern China and a handful of Pacific islands, collectively the fourth largest colonial empire at the turn of the century—were set up for disparate reasons, mirroring the complexity of the colonial debate in the homeland. The German chancellor, Otto von Bismarck, had long opposed the new state's implication in colonial politics, which he thought hid too many geopolitical and financial risks. The Reich's involvement began instead through the activities of private colonization societies roughly modeled on the British East India Company. Only later did state authorities take direct control of any overseas possessions. Whatever the technicalities of colonial control, these territories then needed to be surveyed and mapped to be known and governed. The interest of geographical societies had, since the mid-nineteenth century, been pivotal in the birth and growth of a colonial movement in Germany, promoting scientific expeditions well beyond the areas that later became colonies of the Reich. The reality of Germany's colonial endeavors produced a rather messy situation, in which it was not always possible to set up the necessary infrastructures of measurement.[9]

Different strategies were adopted by colonial officials to handle the multiplicity of possible vertical datums in the various possessions. South-West Africa, Germany's first colony and the only one meant as a settler colony, was also the only one in which a surveying effort was undertaken comparable in its aims, if not in scale, to those of the British in their colonial possessions. Initially, the selected datum was sea level at the tide gauge in the port of Swakopmund, arbitrarily assumed to share the same surface level as the German metropolitan *Normalnull*. To emulate the structure of the German vertical reference network, in 1906 an inland *Normal-Höhenpunkt* was selected. Determined according to a careful leveling along the railway line, itself based on a onetime reading of the Swakopmund tide gauge, this point was set at 1,683.266 meters of elevation in the colony's administrative capital, Windhoek. By literally setting into stone the benchmark, making it independent of sea level in Swakopmund, the colony's vertical datum was put—as had been the case in the homeland—at the geographical, political, and economic center of the land. At the same time, the conventional and political nature of vertical reference points was reiterated.[10]

Rather different decisions were taken in German East Africa (now Tanzania), originally an exclusively commercial enterprise, with minimal surveying needs beyond those required for the construction of railway lines to funnel resources toward the coast. When a vertical datum was suggested there in 1897, the preferred choice was mean sea level in Dar es Salaam. No efforts were made to survey inland for more stable reference points. While sources do not provide clear hints as to what datums were adopted in Germany's other African resource-extraction colonies—Togo and Cameroon—it seems safe to suppose that surveying similarly played a marginal role, linked mostly to the construction of minimal railway networks and the improvement of ports.[11]

In the Pacific, German agents had seized control since 1884 of the northeastern corner of New Guinea and some surrounding islands. In the following years more archipelagos, reaching north to the Northern Marianas and east to Samoa, came under German control. In New Guinea, German rule was concentrated on the coasts, where settlements and plantations were located. Most of the inland remained unknown, and little was done to set up a measurement framework or launch a formal survey of the whole territory; indeed, there did not seem to be a need to formally define a benchmark for heights. When altitudes in the German Pacific colonies were referenced in print, they were generically referred to sea level, without, however, any detail as to how this was recorded. Either elevations were determined barometrically or individual makeshift locations were adopted as datums.[12]

In the last addition to the German colonial empire, the Kiaoutschou Bay concession around Qingdao, the choice of a vertical benchmark was a consequence of the fact that the territory was managed, not by a private company or the foreign office, but by the German navy, whose primary interest lay in keeping the port navigable. Given that and the minimal size of the stronghold, lowest spring tide was adopted as the colony's vertical datum. That this was the only colonial possession it managed fostered the navy's desire to prove its superiority in respect to other ministries, an example of interagency competition. Qingdao thus became a sort of showroom of German modernization, including a scientifically advanced naval observatory, enriched around 1904 with what seems to have been China's first automated tide gauge.[13]

As the German case shows, differences in the standards adopted in the heterogeneous outposts of an imperial power reflected the specific aims of each colonial enterprise and the peculiar roles of the various administrations. Settlement colonies necessitated surveying networks that mirrored those of the homeland, while colonies held primarily for resource extraction, commerce, or military control required less geodetical precision. The limitations of the idea of a standard or uniform vertical datum come to the forefront when considering the colonial dimension. Nonetheless, as discussed in the next section, this same period saw the idea of an international point zero become one of the main aims of the European geodetical community. German officers played a crucial role both in setting up the institutional framework in which this community met and in fostering the debate about standardization itself.

THE DREAM OF AN INTERNATIONAL ZERO

Defining national datums was merely the first step toward standardizing elevation measurements in Europe. The Mitteleuropäische Gradmessungs-Kommission (Central European Arc Measurement Commission) was set up in 1862—before sea-level-based vertical reference points had been defined in most countries. No time could have been better to start working on Europe-wide standardization. As hinted by its name, the commission initially focused on measuring meridian arcs—curves linking points with the same longitude—in Central Europe. This was crucial to determining the local figure of the earth and ascertaining how it differed from that of neighboring regions. If the overall geometrical shape of the planet had been determined as accurately as possible in the previous decades based on various arc measurements by Airy and Bessel, more work seemed necessary to explain local distortions. In particular, the commission's founders lamented the lack, despite the presence of a thorough network of triangulations, of an arc measurement in Central

Europe. The importance of vertical datums became immediately apparent, especially in view of the theoretical move toward modeling what the German mathematician and geodesist Carl Friedrich Gauss called the earth's "true (geometrical) shape," an idealization of its equipotential surface at sea level. Accordingly, the link between mean sea level, arc measurement, and trigonometric altimetry was stressed repeatedly at the conferences organized over the following decades, during which the commission made the definition of a unitary pan-European point zero of elevations one of its core tasks.[14]

This intergovernmental organization was born out of an idea of the Prussian military surveyor Johann Jakob Baeyer, a longtime friend of Bessel, to promote the measurement of a meridian arc between Palermo and Oslo. At the beginning it included representatives of Belgium, the Netherlands, Switzerland, Italy, Austria, Poland, Denmark, Sweden, and Norway, as well as Prussia and several other German states. France at first declined membership, but did make its geodetic data available. Over the years new countries were admitted, and the commission was accordingly renamed: in 1867 it became the Europäische Gradmessung (European Arc Measurement), in 1886, the Internationale Erdmessung (International Geodetic Association).[15] By 1889 some extra-European countries—the United States, Mexico, and Japan—had joined, ratifying the global scope implied in the new name.

The United Kingdom was the last major country to join, in 1898. In the first report he presented to the association, the British astronomer George Darwin, son of Charles Darwin, justified the delay, explaining that the British Ordnance Survey, on the one hand, was satisfied with the precision already attained by the principal triangulation concluded in 1852 and, on the other, had possessed, in the intervening decades, only enough financial resources to focus on the government's demand for large-scale, detailed maps. Moreover, as Great Britain's colonial conquests expanded, British scientists lived and worked under the impression that they could perform most of the same tasks as the international commission within the bounds of the empire, and possibly perform them better. After World War I, the commission, purged of its German roots, was refounded as the Geodetic Section of the International Union of Geodesy and Geophysics (IUGG), which then became the International Association of Geodesy (IAG).[16]

The association, serving as both an information exchange office and a space of scientific debate, played a crucial role in consolidating mean sea level as the global reference point of choice for altitudes. Early in its history, it was decided that the definition of a common European point zero of elevations should be based on careful research and thorough surveying—a response to the perceived lack of studies supporting the French adoption of mean sea level at Marseille as a national datum. The search for a continental reference point for elevations was therefore structured as a long-term endeavor, requiring the collection of longer observation series and improvements to the geodetic infrastructure. Nonetheless, hopes were high that a common benchmark could be defined rather swiftly. Numerous issues, however, from the need to explain the measured discrepancy in height among different European seas to the difficulty of convincing states to abandon their national datums, slowed the process, and decades of debate brought little progress. A growing number of national representatives even started to question the usefulness of a common reference point, when locally determined mean sea levels could easily be compared and converted from one framework to another.

Some of the challenges were structural. The ability of the commission, as an intergovernmental as well as a scientific association, to reach consensus depended not only on the scientific soundness of its decisions and compromises, but on how well they reflected the interests of all member states. Leading the association, through all its incarnations, was a permanent select commission, which met annually, charged with coordinating the group's scientific activities. It was assisted by a Berlin-based central bureau (from 1870 onward, embedded in the new Prussian Geodetic Institute, another product of Baeyer's initiative), which collated reports about member countries' geodetical activities, managed the commission's archives and library, and published an annual report. Every three years the association also organized a general conference, where representatives of all member countries met. The German states played a pivotal role: not only was the central bureau essentially an office of the Prussian government, but all general conferences until 1880 took place on German or Austrian territory. Moreover, until 1886, well after the institution of the unitary Reich, most major German states were full members of the association and sent their own representatives to its general conferences. Consequently, German representatives were also a plurality on the permanent commission.[17]

Following the debate through about three decades of meetings may show more vividly the complex interactions among participating scientists and practitioners. These were at the same time representatives of national interests and individuals with their own scientific beliefs. At the association's first conference, held in Berlin in 1864, the Swiss representative, the astronomer and geophysicist Adolph Hirsch, lamented the lack of studies supporting France's adoption of the mean level of the Mediterranean in Marseille as its vertical datum. The choice of a Europe-wide benchmark, he insisted, should be grounded on thorough research: first, each member country should choose its national reference point; next, these points should be connected by leveling and the instruments compared; finally, mean sea level should be ascertained in as many European ports as possible and the zeroes of the gauges linked to the leveling network. Only then should a European point zero of elevations be defined. Other representatives enthusiastically concurred. The Berlin-based meteorologist Heinrich Dove, worried about the quality of the usual mean sea-level observations, further recommended that self-registering tide gauges be used.[18]

At the second conference, in 1867, Wilhelm Förster, the director of the Berlin Observatory, suggested that mean sea level be measured where it was easiest. Hirsch responded, stressing the importance of measuring at multiple locations:

> This theoretical mean sea level can't be determined experimentally in respect to the surrounding continents other than by setting up systematic observations of the elevation of the seas and continents at as many points as possible and consequently of the relative elevation of the different seas across continents. The mean level of one sea is an abstract measure and can, just as mean air pressure and mean temperature, be determined only from a longer series of observations of the variables possibly affecting the level of the sea.[19]

Hirsch was reiterating that there weren't yet enough data available to settle upon a common European datum. More work would have to be done to determine which location on what sea would best serve this purpose. Defining vertical datums based on local mean sea levels would contribute to assessing the viability of candidates for a pan-European reference point, but it was, as he'd said three years earlier, just a first step.

At the 1871 conference, for example, the Italian representative, the geodesist Federico Schiavoni, recommended that the association's permanent commission ask the Italian government to install self-registering gauges on the Tyrrhenian Sea in Paola, on the Ionian in Rossano and

Taranto, and on the Adriatic in Brindisi. All four stations should then be connected to the leveling network to determine whether the three seas were part of the same equipotential surface. Thereafter, their relative elevation should be reassessed every ten years.

The technical improvement of self-registering gauges provoked great expectations in the geodetical community. According to Baeyer's daughter, mareographs were his greatest delight. Hopes remained high that the choice of a common reference point would be relatively swift and easy.[20] Yet the data produced by national survey efforts recorded differences in level among the seas surrounding Europe. Already Bourdalouë's work, started in 1857, had registered a discrepancy between the Mediterranean Sea and the Atlantic Ocean of more than a meter. When in the early 1870s German surveyors connected the leveling of Alsace—recently annexed from France and thus referring to mean sea level in Marseille—to the tide gauge in Swinemünde, they recorded a difference of 0.74 meters between the Mediterranean and the Baltic Sea.[21]

In 1879, at the meeting of the permanent commission in Geneva, president Carlos Ibáñez e Ibáñez de Ibero, the Marquis of Mulhacén—described by the American philosopher and geodesist Charles Sanders Peirce as "more a man of the world than a man of science"—volunteered to report on the diffusion of mareographs along Europe's coasts. At subsequent general conferences, Ibáñez dutifully presented his reports, repeatedly urging the member states to connect by leveling all working tide gauges. In 1883, at the Rome conference, Hirsch noted how, despite the efforts of individual nations, the process of unifying heights in Europe had made little progress. National leveling endeavors had not yet been interconnected, and the data provided by tide gauges were still insufficient. Furthermore, Hirsch noted, the average difference in level between the Atlantic and the Mediterranean registered by Bourdalouë coincided approximately with the one detected in repeated surveys between Santander, on the Bay of Biscay, and Alicante, on the Mediterranean. A debate ensued on the relation of observation errors to actual differences in level. The discrepancies recorded so far were beyond the statistical margin of error. Attempts to attribute them to differences in salinity and density among the various seas did not fit with the actual data from the surveys. Baeyer had suggested already in 1875 an alternative hypothesis: that, over long distances, such measured discrepancies might arise from a north deflection of the plumb line due to increased gravitational attraction over the continent.[22]

In 1886, at the eighth general conference in Berlin, Charles Lallemand, head of the French Service du Nivellement General, the office in charge of the country's national leveling effort, read a theoretical paper that took up a point recalling Baeyer's theory. Lallemand suggested that because of the varying influence of gravity, the earth's surface levels weren't parallel, and this might lie at the root of a systemic error in all previous transcontinental levelings. He theorized that "one would find as many values for the difference in level of two points as there are paths to go from one to the other"—a nifty explanation for the discrepancies recorded between the elevations of the seas in repeated levelings. To test this, he had commanded the office he led to systematically correct levelings for gravity, in what he called "dynamic leveling." Hirsch, the association's secretary, remarked how much more work in the field such adjustments would entail and doubted they would lead to any significant improvement in the results.[23]

At the 1888 meeting of the permanent commission in Salzburg, Hirsch read a proposal reaffirming the commission's commitment to select a common European point zero of elevation. Commission members agreed that such a determination not only would be useful for those countries that had not yet selected a vertical datum, but would make it easier to convert heights between differing reference frameworks. This would benefit engineers working on cross-border infrastructure projects, such as railways or canals. Appealing to both scientific and political reasons, the commission suggested that the international point zero be located in a North Sea port that did not belong to a major continental power.

As soon as president Ibáñez opened the discussion, Lallemand, who, though not a member of the permanent commission, regularly attended its meetings, was the first to take the floor. He went to the core of the issue and questioned the usefulness of the whole endeavor: since most nations had selected datums, converting among them simply meant adding a constant. The choice of a European point zero, for which more research seemed necessary, might actually lead to more errors. Moreover, Lallemand worried, if a North Sea port were chosen, negative heights would be recorded along the coasts of the Mediterranean Sea. Hirsch dismissed Lallemand's criticisms as insufficient to delay the choice of a common European reference framework for elevations. Others, like the French hydrographer Anatole Bouquet de la Grye and Annibale Ferrero, the director of the Italian Military Geographical Institute, chose a middle ground; while they saw no reason not to start thinking about how to determine the European datum, the commission should "allow the necessary time to give plenty of thought to the choice of the best zero point"—which, contrary to the permanent commission's proposal, would best be chosen, they claimed, where the difference between high and low tide was the smallest: that is, the Mediterranean. National preferences and scientific localism played an important role in this debate, and each country, and even regions within the major countries, pushed for its local sea level to be used as the European common reference point.[24]

ONE LEVEL FITS ALL

The debate over an international point zero would continue for at least another decade. Soon, however, new research showed that earlier surveys had vastly overestimated the differences in elevation between European seas. According to new, more accurate levelings, recorded discrepancies were actually well within the statistical margin of error. This finding strengthened the idea that any given sea level was just the local instance of a common level. In the wake of Lallemand's critique, an increasing number of representatives began to ask what purpose a fixed common benchmark would serve. A single international point zero seemed a transitory achievement that would need to be continually reassessed. The idea of sticking with national datums and conversion tables gained traction, especially among the representatives of major countries. While no explicit decision was ever made to stop doing research on the issue, the search for a common European reference pointfinally dwindled into oblivion, effectively disappearing from the publications of the Internationale Erdmessung.

At the general conference in Paris in 1889, the year after he first questioned the usefulness of a European vertical datum, Lallemand reported on the activities of the Service du Nivellement General. New measurements, he said, showed the difference between the mean levels of the Mediterranean Sea and the Atlantic Ocean to be much less significant than previously thought, amounting to just about 20 centimeters. Bourdalouë's levelings, which for three decades induced the geodetical community to overestimate the difference, had been flawed by systemic errors. And the Spanish leveling of 1883 had neglected to consider how the influence of gravitation varied with changes in latitude.

In view of these and similar results showing essential uniformity in mean sea levels, and considering the precision of existing instruments,

Lallemand suggested that the unification of the datum had in practice been accomplished, insofar as most national geodetic frameworks already referred to some local mean sea level. In the public discussion, held on October 9, Ibáñez reiterated the stance of the permanent commission, that further research was needed. Work should continue, he said, ahead of the following general conference, when a decision would finally be taken. Ferrero, believing time was needed to assess all the involved variables regarding characteristics of sea and coast, was particularly pleased with this decision. Given the effect of distance on the accuracy of leveling, he also wondered whether it might not be preferable to abandon the idea of a coastal reference point and pick instead a location in the middle of the continent. Hervé Faye wondered whether adopting immediately Ferrero's proposal might allow the commission to set a preliminary standard, which could later easily be connected to the chosen sea.[25]

In a paper presented at the 1890 meeting of the permanent commission in Freiburg im Breisgau, Lallemand attempted to strengthen his theoretical stance. Recent geodetical work had confirmed, he said, that differences in level among the seas encircling Europe were within the surveyors' margin of error. (Alexander von Kalmár, the Austrian commissioner, had independently reached the same conclusion.) If each maritime country, then, picked as datum a fixed point along one of its coasts, the unification of heights throughout Europe would be obtained, with no challenge to national pride or need for complicated compensatory calculations to align the various leveling networks. Indeed, if Germany and the Netherlands lowered their reference points by about 15 to 20 centimeters, so as to reflect the actual mean level of the North Sea, the unification of European datums would already be a reality: all of the most important networks would then refer to mean sea level. Lallemand was convinced that the effort needed to redo the many compensation calculations would exceed the benefits of choosing an arbitrary common European point zero of elevations, which, in the best of cases, would modify elevations throughout Europe by no more than 20 centimeters.[26]

To fix the relation between European hypsometry and the levels of the seas, claimed Lallemand, would "only be an illusion, the satisfaction of a moment." Scientific and technical progress would continually improve the exactness of surveys and the accuracy of the leveling network, forcing European countries to repeatedly revise recorded altitudes and existing landmarks. If, for instance, the Amsterdam datum had been chosen as the international reference point before France corrected its hypsometric network, compensating for the errors of the earlier Bourdalouë leveling would have required that Spain increase all of its recorded elevations by about 80 centimeters. Insistence on a unitary reference point, held Lallemand, would have the paradoxical consequence of increasing instability. The scientific community would encounter problems similar to those it would have faced had the original definition of the meter as a fraction of the meridian arc been preserved: with each new, more precise measurement of the meridian arc, the basic unit of the metric system would have had to be changed.[27]

At the following meeting of the permanent commission, the Prussian representative and head of the association's central bureau, Friedrich Robert Helmert, also renounced the idea of a single European datum. The Internationale Erdmessung published the differences between member countries national reference points, and this, he claimed, would suffice for all practical needs. Hirsch, the secretary, remained unconvinced and stated that the discrepancies in sea elevation across Europe represented only a secondary issue. The real issue concerned the differences recorded at national borders, which he thought could reach several meters and present major problems for railway and canal engineers. These issues, he believed, could only be overcome by selecting a common zero for the whole continent. Lallemand countered:

> If mean sea level in a given location was chosen today as the international point zero of elevations, the transfer of this level between the different countries would have been achieved by the sea itself with as much, if not more, precision than through the mediation of the existing levelings.

Nonetheless, the permanent commission agreed with Hirsch that a final decision should be deferred to the next year's general conference; in the meantime, the central bureau should continue to support research on the matter.[28]

Helmert assigned the task of reporting on advancements in the field to the expert geodesist Anton Börsch. In his report, discussed at the tenth general conference of 1892, Börsch noted, among other things, that there was no significant variation among mean sea levels measured along the coasts of each individual sea or among those recorded for different seas. He further agreed with Lallemand that completing the compensation calculations to accommodate the European height network to a

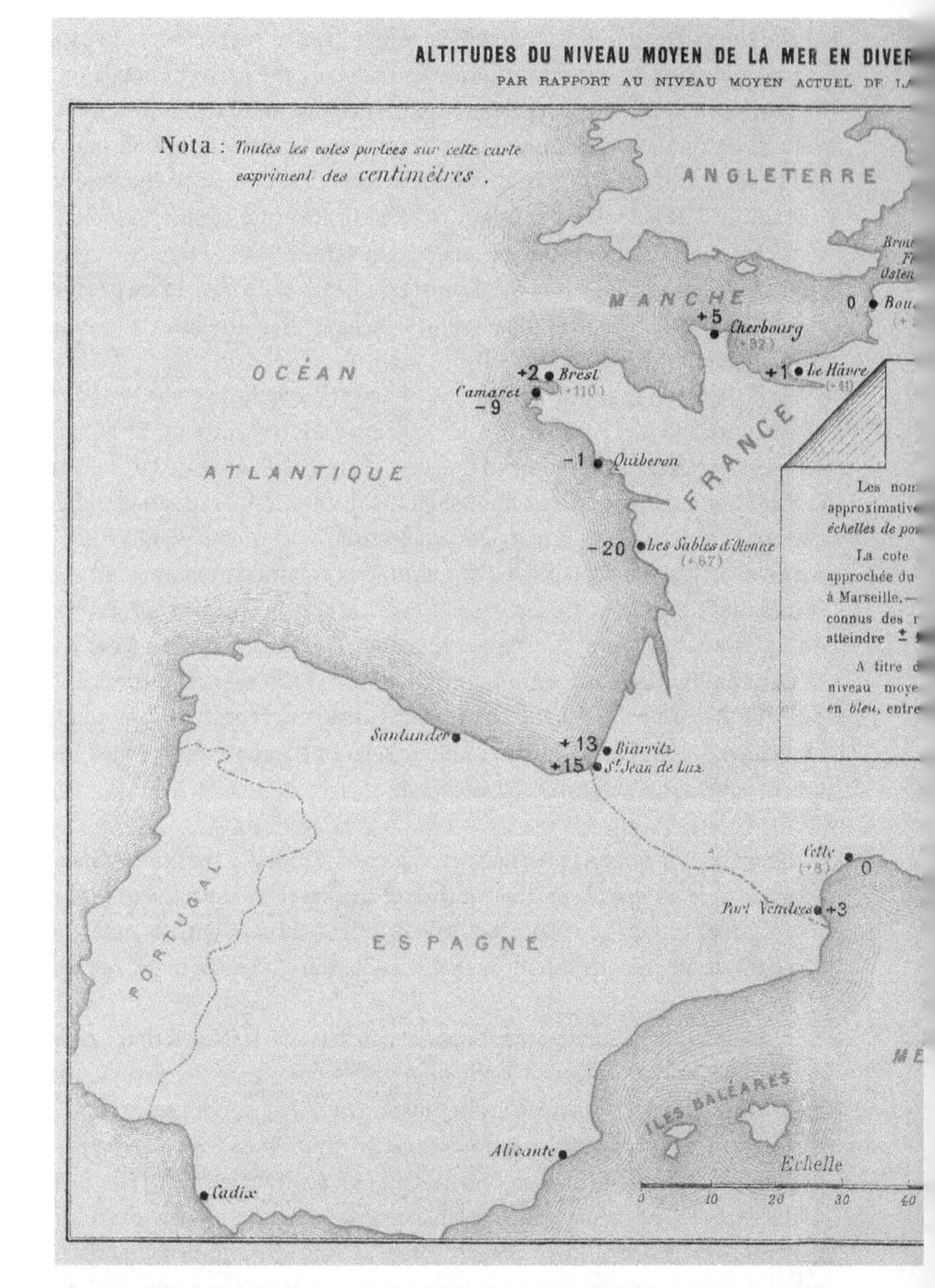

FIGURE 3.2. The central offices of the Internationale Erdmessung produced tabular data and cartographic material intended to clarify how one might define an international point zero of elevation. This map, for instance, illustrates the differences between mean sea levels in Marseille and at other European ports. *Source:* Hirsch, *Verhandlungen der . . . Conferenz der Permanenten Commission*, 1891, plate IIa.

…OINTS DES CÔTES EUROPÉENNES

…DITERRANÉE, A MARSEILLE

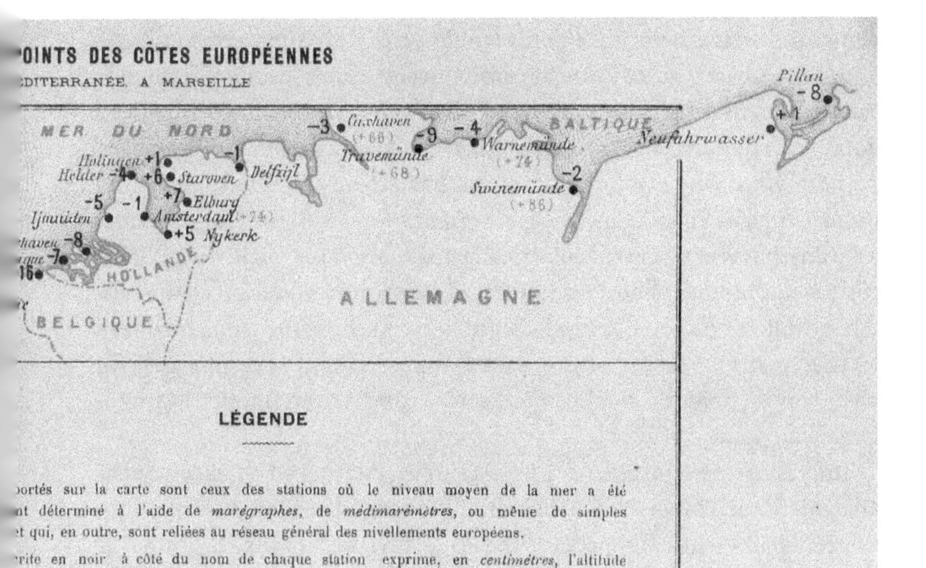

LÉGENDE

…ortés sur la carte sont ceux des stations où le niveau moyen de la mer a été …nt déterminé à l'aide de *marégraphes*, de *médimarémètres*, ou même de simples …t qui, en outre, sont reliées au réseau général des nivellements européens.

…rite en noir à côté du nom de chaque station exprime, en *centimètres*, l'altitude …eau moyen dans cette station, par rapport au niveau moyen actuel de la Méditerranée …s altitudes ont été calculées en soudant, après une compensation sommaire, les résultats …llements de l'Europe Centrale et Occidentale. L'erreur probable des cotes obtenues peut …centimètres.

…comparaison, on a indiqué, pour quelques stations, l'altitude anciennement admise du … par rapport au même niveau de comparaison. **Ces cotes rétrospectives sont écrites …arenthèses, à côté des noms des stations correspondantes.**

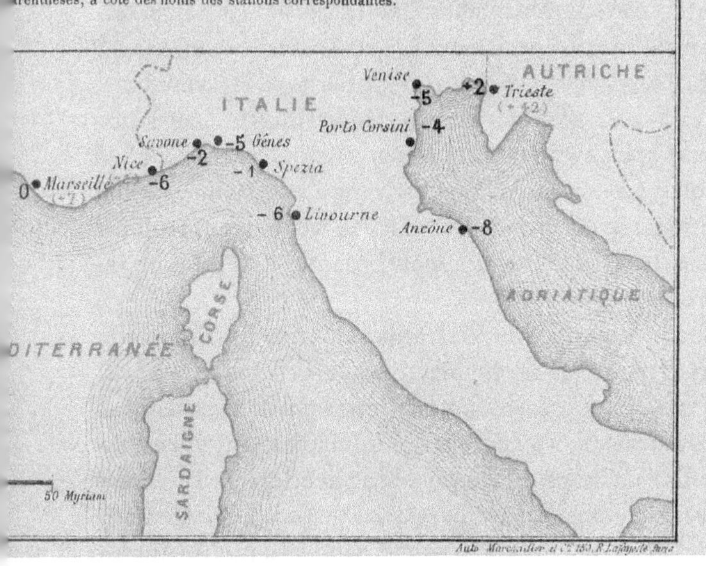

Auto. Marchadier et Cie 150, R. Lafayette Paris

common vertical reference point would be unduly time-consuming—and they would have to be redone after every future leveling, impeding the advancement of science. Even if all of the expenses and difficulties entailed in pursuit of a moment of unity made sense from the engineers' point of view, why should nations that had already selected a reference point—such as Prussia and the Netherlands—redo all the work to adopt a datum that was, at best, scientifically disputable? In conclusion, Börsch reported, the central bureau reiterated its recommendation against establishing a common European point zero. At the same time, however, he used in his tables the Amsterdam datum as a point of comparison for all European height reference points, elevating it to an implicit status as European point zero.[29]

Italy's representative, Annibale Ferrero, countered that, while the analysis Börsch presented on behalf of the central bureau carefully considered mathematical aspects of the issue, it was otherwise lacking. The choice of a common benchmark, he claimed, was not merely an issue of precision leveling. It involved bigger, practical questions, such as the need to determine, first, the inherent hydrological and geological features that a vertical datum should have and, then, whether there were one or more points in Europe that might satisfy those requirements. Ferrero found the central bureau's decision to drop the search because of the limited precision of levelings premature, to say the least. He feared that political caution and personal preferences were diverting the association from its scientific duty to investigate a question many member states had posed. In particular, he wondered—highlighting a question pertinent to much international scientific cooperation—to what extent national considerations precluded a solution, simply because "that solution involuntarily favored one country."[30]

Hirsch wholeheartedly supported Ferrero's motion to appoint a commission of experts to continue the inquiry. It would be a shame, he said, to quit, after almost thirty years, without reaching some consensus on an issue of crucial practical importance to major infrastructural endeavors. The choice of a common European zero would not, he acknowledged, have any consequence for countries that, "for reasons that are fundamentally foreign to science," preferred to preserve their national datums—the association had no formal authority to impose its decisions on anybody. It would, however, allow many other countries to determine the "absolute" values of elevation within their boundaries. All major countries had already selected datums, and in doing so had confronted the problems that the central bureau now concluded made an international point zero useless and unfeasible. On the national scale, as on the international, errors accumulate over great distances. Lest any nation's pride be threatened, Hirsch proposed as a common European reference point, not a measurement tied to a single port, but the average of Europe's mean sea levels, related to and marked at a stable point in a neutral country close to the geographical center of the continent. Consistent with Hirsch's idea, his home country of Switzerland would in 1903 correct the recorded elevation of its datum on the Pierre du Niton in the port of Geneva to reflect the averaged and weighted mean sea levels of adjoining countries.[31]

Helmert countered that it would not be in the interest of the association if the zero it proposed had soon to be abandoned because of advances in leveling techniques. Future advances might allow the definition of a common level for the whole of Europe. But not anytime soon: the continent was just too large. All in all, however, Helmert saw no need to cancel the project outright, and did not oppose a further postponement of any decision. He thus agreed to the appointment of a new commission—composed by Hirsch (Switzerland), Kalmár (Austria-Hungary), August Morsbach (Prussia), Gerrit van Diesen (Netherlands), and Lallemand (France)—tasked with specifying what conditions a European point zero should satisfy and determining which locations were most apt.[32]

The following year, at the 1893 meeting of the permanent commission in Geneva, Lallemand reported on the commission's preliminary work. An ideal international zero likely did not exist, remarked the Austrian representative, Kalmár: given the constant variability of relative sea level and the impact of subsidence on inland locations, no single location seemed to fulfill the criteria that a datum be both centrally located and invariable in time and space. He concurred with the Dutch representative, van Diesen, that mean sea level should be adopted. But while the latter believed that a series of local measurements could suffice, Kalmár agreed with Helmert that the exact position of the equipotential surface should be determined, which was not possible until the accuracy of measurement techniques improved. Lallemand suggested therefore, once more, that each country use the closest mean sea level as datum. Since all seas have approximately the same altitude, the unification of the hypsometric networks would, he posited, be obtained without further effort.[33]

The results of the commission's inquiry were expected to be divulged at the 1895 general conference, but no final report ever appeared. It is possible that, to avert further politically charged debate, the members agreed to disagree and quietly forget about the issue. Those who had sought to define a common zero would avoid a public debacle. Their antagonists had nothing to lose. The status quo favored them: if no decision was taken, each country would keep its national datum, and, according to Lallemand's insight, the unification of the hypsometric network would be accomplished, with no need for arbitrary compensation or new leveling. In the following decades the issue was presented in the specialized press as substantially resolved. And the idea that a centrally located datum would have served better than a coastal one was soon dismissed: the perceived long-term stability of the oceans seemed preferable to the risk that landmarks and reference points might be displaced by movements of the earth's crust.[34]

After years of fierce scientific debate, the attempt to determine a Europe-wide vertical point zero just petered out. Even so, the debates changed the way datums were understood. Comparing how vertical reference points were chosen in different European countries and tracing the history of the search for an international point zero reveals a steadily growing acceptance of the conventional and customary nature of datums. Year after year more and more geodesists accepted the fact that the tridimensional grids they were setting up to measure the earth were useful abstractions, not reflections of a "natural order." Connecting sea-level data produced along Europe's coasts with the ellipsoid—a geometrical approximation of the planet's shape, whose regularity eases cartographers' work—was most easily achieved on smaller spatial scales. A national dimension, too, came into play, for both political and practical reasons, independent of a nation's dimensions. Multiple overlapping ellipsoids, each passing through the local vertical datum, are thus used in different national contexts to minimize the difference between the actual shape of the planet and its geometrical approximation.

While the push to define national datums had been apparent since 1864, the debate over a European zero gained momentum only in 1879. The process by which datums are chosen sets in place distinct path-dependencies. Once mean sea level in the port of Marseille had been chosen as the French national reference point, considerations of cost and efficiency prevented changing it, even as the arbitrariness of the choice became increasingly clear. In Austria, soon after the mean level of the Adriatic in Trieste was selected as a datum, surveyors realized that, because of the short period in which data had been gathered, the given level was only accurate within 1 or 2 centimeters; the datum was then redefined as an invariable distance below the landmark on the Molo Sartorio, rather than the constantly updated, corrected mean sea level.[35]

Germany selected its datum later than France and Austria, and so could learn from their examples. Developments in Germany paralleled the acceptance by Austrian geodesists that choosing a reference point that would remain stable in the long term was, practically and politically, more important than absolute scientific fidelity to the measured mean sea level. Technical advances could not warrant continually changing the datum. The French adopted a similar stance when the new mareograph in Marseille began producing longer data series: while the original reference point had been chosen almost on a whim, the costs of change contributed to its preservation. From early on, their German counterparts acknowledged that the mean level of the sea is not, intrinsically, more or less significant than the level of any other arbitrarily chosen point. Lallemand—the standard bearer for the idea that all vertical datums should be linked to mean sea level—saw the emphasis placed on uniformity by supporters of an international zero as a matter not just of defining new, more effective methods to read the world, but of complete abstraction. The units of measure and the frameworks of reference were moving farther and farther away from the object they should represent.[36]

In the end, the representatives of the German and French points of view—Helmert and Lallemand—agreed on one thing. In a system based on conventions and faced with problems of instrumental accuracy and computational precision, the definition of an international point zero was superfluous. Their stances in a debate that had increasingly become a political and diplomatic confrontation were much closer than one would expect. Locally measured mean sea level remained the most widely adopted reference point. It was accepted, however, that it was a rough proxy when not a mere abstraction, a fact that furthered its epistemic shift from "natural object" to symbol. As a consequence, the material connection of the vertical datum to the sea became ever less relevant.[37]

FOUR

THEORIES OF CHANGE

Eggenburg is a small town in Lower Austria. It lies on the banks of the river Schmida, a northern tributary of the Danube, on the fringes of Austria's wine region and at the heart of the lowlands extending in front of the Alps. Even if the town sits 360 kilometers from the nearest sea, at Trieste in Italy, the landscape that surrounds it played a crucial role in renewing scientific acceptance of the idea that sea level can and does vary. In the early 1860s, while hiking in the region to research the shores of an ancient sea, the Austrian geologist Eduard Suess noticed ample, homogeneous sea deposits along the slopes of the gently rolling hills. Recalling similar formations at the margins of the Hungarian plains, about 250 kilometers to the southeast, he considered the idea that uniform deposits over such a distance could hardly be due to a synchronous elevation of the land. Contrary to dominant geological theories, Suess began to think, only a drop in sea level could cause such formations. This reinterpretation of land-formation—decades before the development of plate tectonics—had broad repercussions, including, as Suess himself suggested, the potential to explain biogeographical oddities like the existence of closely related animal and plant populations on islands and continents separated by the sea. Much had still to be explained and measured, as Suess would recognize in a posthumously published memoir. Not till fifteen years after those initial Austrian hikes would he take this stance publicly, but the seed of a theory allowing for the possibility of absolute, global sea-level changes (so-called eustasy)—not just apparent ones due to local movements of the land—had been sown.[1]

In recent years the debate over labeling the current geological era the Anthropocene, the age of man, has brought with it a revival of catastrophism—an understanding of how change occurs in geological processes that favors the role of sudden, violent events. Catastrophism had apparently lost its intellectual battle against gradualism and uniformitarianism in the course of the nineteenth century. But anthropogenic changes have ended the stability of the holocenic "long summer" in which early geological thinking developed. The idea of a long-term uniformity in geological processes, proposed and defended by the likes of James Hutton and Charles Lyell, has been disrupted.

If Leopold von Buch's and John Playfair's take on a stable sea facing a moving land was the most widely accepted explanation for changes in their relative position through most of the nineteenth century, there was, in the decades preceding Suess's insights, no lack of alternative theories. The presumption that local movements of the land alone explained changes in the relative height of the sea had increasingly been questioned and, indeed, seemed at odds with the precepts of Lyellian uniformitarianism. Over the course of the century after ice ages were first theorized—in the late 1830s—they came to play a crucial role in answering a question that had stymied previous theories conjecturing cyclical changes in sea level: assuming that the overall volume of water on the planet is essentially stable, where does it go when the oceans' level decreases? Glaciation did not immediately become the dominant theory. Other, competing theories minimized the role of glacial processes, but they too helped give the idea of an unstable sea renewed credibility in scientific circles, and they broadened the overall understanding of planetary mechanics. These alternative theories produced new ways to comprehend the relative position of land and sea and gave rise to a whole new subdiscipline, geomorphology, dedicated to the processes behind land formation. The ongoing debate among geologists about the stability or instability of sea level also influenced the work of geodesists in the early decades of the twentieth century and made possible the development of a new theory consolidating the role of glaciation in sea-level variations on both geological and historical timescales. The interconnectedness of sea-

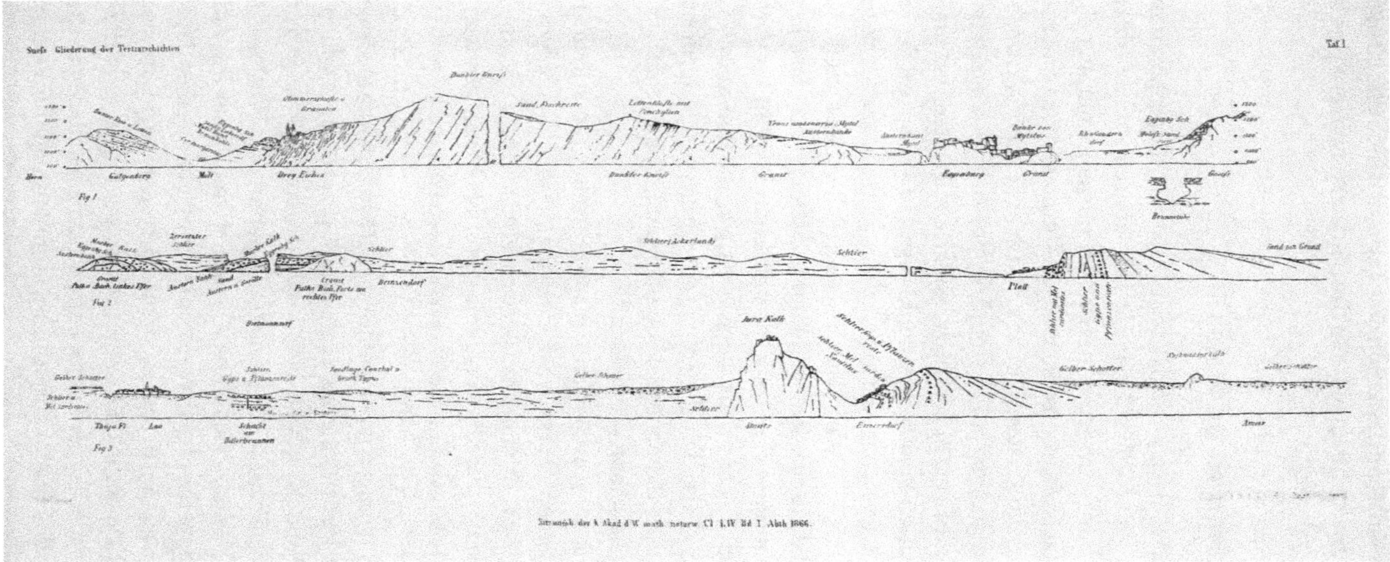

FIGURE 4.1. The peculiar geographic profile of the landscape around Eggenburg, Austria, and its resemblance to that of other hilly regions, induced Eduard Suess to theorize absolute sea-level changes in geological times as a central component of landscape formation. *Source:* Suess, "Untersuchungen über den Charakter der österreichischen Tertiärablagerung," plate 1.

level variations and the alternation of ice ages and interglacial periods is exemplary of how major changes in the planetary environment need to be understood holistically.[2]

FROM GLACIERS TO BEACHES

In 1840, the Swiss-born geologist Louis Agassiz first proposed in print his theory that glaciation had played a primary role in the formation of landscapes. Just two years later, the Scottish journalist and amateur geologist Charles Maclaren posited the potentially momentous consequences of such an ice age on sea levels. In an extensive review of Agassiz's work, Maclaren noted that glaciation would, by necessity, lower sea levels by freezing large amounts of water on land. He estimated that glaciation of just the northern hemisphere could result in a sea-level decrease of one to two hundred meters. In the following years, others began to think along similar lines, rarely, however, making their stances public. In 1844, for instance, the French geologist Eugène Robert came close to reconsidering the root cause of elevated ancient shorelines, in a report to the Académie des Sciences on his work as part of the Commission scientifique du Nord, charged by the French government to perform research on Scandinavia. He did not dare, however, publicly contradict received wisdom about the stability of the sea.[3]

In 1842, the same year Maclaren made his timely remark, the French mathematician Joseph Adhémar claimed that it was commonly accepted in geology that the continents had been covered multiple times by water. And indeed, Lyell included in later editions of his *Principles* a map showing the extent to which Europe may have been covered by water at one time or another. Lyell found the causes for these massive floods in gradual and recurrent crustal movements rather than absolute changes in sea level. Adhémar instead offered a catastrophist reading, in which floods were a regular and inevitable, if sudden, occurrence, caused by a combination of astronomical and geological phenomena. While his theory was criticized by many, including Alexander von Humboldt, Adhémar helped consolidate the idea that glaciation might be an essential component of land-formation processes in the current geological period—the Quaternary, which began 2.58 million years ago. The limits of the theory notwithstanding, Adhémar's work enjoyed some popularity in the years immediately following its publication, stimulating multiple responses and being translated into German as early as 1843.[4]

Not trained as a geologist, Adhémar treated the issue of glaciation cycles as an exquisite mathematical problem to be analyzed and solved at a purely theoretical level. His main tenet was that the center of gravity of the world's oceans moved over time following the expansion of the ice cover in one or the other hemisphere. In the course of such a shift, he imagined, water would slosh from one hemisphere to the other, causing a fall of sea levels in one and the submersion of continents in the other. The causal explanation he gave for these shifts lay in the precession of the planet's rotational axis. According to Adhémar, these changes, influencing the length of polar nights, alternately affected the temperatures of the hemispheres, on a regular and predictable cycle of 10,500 years. The increase in polar ice caused by the cooling of one hemisphere would increase as well the volume of liquid water attracted toward that pole and thus prompt a rise in local sea level.[5]

But not all work on sea-level changes at the time was purely theoretical. In 1849, as part of a broader expansion of transportation networks, a new railway line was inaugurated, running from Edinburgh to Hawick in the Scottish Borders region. During its construction, excavation of the foot of Buckholm Hill, near the town of Galashiels, revealed an alternation of hill debris and marine sand at an elevation of about 350 feet (100 meters) above sea level. Robert Chambers—a Scottish publisher and amateur geologist whose measurements of ancient shorelines near St. Andrews in Scotland had already gained him membership in the Geological Society of London—visited the site and was inspired to theorize multiple significant sea-level rises and falls as the prime motor of land formation along the coasts of Britain. In his 1848 book on the role of "ancient beaches" as markers of long-term changes in sea level, Chambers cited examples from Scotland, the rest of Britain, Europe, and North America, showing how geological remnants of multiple, sequential changes in sea level could be detected in the soil and the landscape. Past stable states of the sea remain evident in the geological record as alternating layers of fossil marine shells and peat, while regular elevations shaped like tidal slopes act as topographical markers. In the following decades, Chambers's work would be cited as seminal by many scholars who favored alterations in sea level as the driving force in changing the shape of the land and coastlines.[6]

The fact that such geological features could be found all over Britain at similar altitudes, led Chambers to assume that all must have the same cause, which must necessarily have been the sea itself. The suggestion ran counter to Lyell's preference for local movements of the land as the main cause of changes in relative sea level. But localized vertical shifts, Chambers argued, could not have produced the uniformity evident in his numerous examples. On the lookout for a cause for global absolute changes, he invoked the subsidence of distant ocean beds suggested by Charles Darwin in his theory of coral formation. Chambers also offered evidence that the latest sea-level shift might have occurred after humans settled the flatland between Perth and Dundee. This included toponymic hints, the legal granting of salmon fishing rights in an area distant from the coast, and archaeological findings of ancient anchors, boat-hooks, and even harpooned whales. While many of Chambers's proofs may well have been linked instead to glacial movements, they hint at an early recognition that sea-level changes might have occurred not only on a geological timescale of hundreds of thousands or millions of years, but also a historical one of mere centuries.[7]

CHANGE AS AN ONGOING PROCESS

In 1853, the English geologist Alfred Tylor voiced his own criticism of Lyell's stable sea theory, then at the peak of its acceptance. Dismissing one of the main tenets of most geological and geodetic work at the time, he denied that the sea may be considered stationary "for practical geological purposes." Renewing older debates, he asserted that sea-level change could be an ongoing process, not just a historical and geological one. Citing a number of eighteenth-century studies, he stated that silt and sediment brought into the oceans by rivers necessarily displaced an equal volume of water, leading to a gradual rise in sea level. To the claim that such additions were negligible, he responded:

> The mere consideration of the number of cubic feet of detritus annually removed from any tract of land by its rivers, does not produce so striking an impression upon the mind as the statement of how much the *mean surface level* of the district in question would be reduced by such a removal.[8]

As examples he gave data collected by, of all people, Lyell, according to which the Ganges river would lower its drainage area by a foot (about 30 centimeters) over 1,751 years, while the Mississippi would take 9,000 years to do so. Tylor hypothesized that the alluvium from all of earth's rivers would be enough to raise global sea level by 3 to 4 inches (about

10 centimeters) over ten thousand years. Tylor admittedly lacked proof that any such fluctuation had occurred, since rising water hides its former levels: in contrast to the ancient beaches left by falling sea levels, shorelines and other markers that preceded a rise would be lost to the waters.[9] In 1868, Tylor revisited his estimates in a paper he read at the Geological Society of London, with Lyell in attendance. Tylor suggested that a generalized fall in sea level caused by the sequestration of water in the form of ice might explain the formation of Pacific coral reefs as well as the massive oceanic subsidence proposed by Darwin. Like Chambers, and later Suess, Tylor held that variation in sea level was the only mechanism that could explain uniform changes over vast geographic areas: land movements would cause evident distortions in the aspect of geological layers, and the extent of their effects would be more limited. The contest over the primary factor in large-scale land-formation changes—land movement versus sea-level change—took place within the boundaries of Lyellian uniformitarianism. There was no place in the scientific debate held at the Geological Society for catastrophic, sudden changes, such as those posited by Adhémar. What was up for discussion was not whether massive changes had occurred, but what slow, uniform cause best accounted for them.[10]

At the 1867 meeting of the American Association for the Advancement of Science, the Ohio-based geologist and surveyor Charles Whittlesey presented a paper on the impact of glaciation on the level of the oceans. While the paper was not much echoed in later works, Whittlesey offered

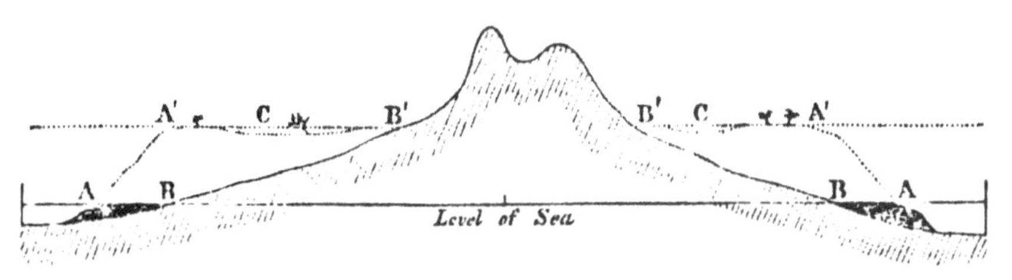

FIGURE 4.2. Absolute sea-level variations were proposed as early as the 1860s as a possible cause for recorded changes in the elevation of Pacific coral reefs. The image shows the coral reef–formation process according to Charles Darwin, with subsidence represented by an apparent rise in sea level: "A A–Outer edge of the reef at the level of the sea. B B–Shores of the island. A′ A′–Outer edge of the reef after its upward growth during a period of subsidence. C C–The lagoon-channel between the reef and the shores of the now encircled land. B′ B′–The shores of the encircled island." *Source:* Darwin, *Geological Observations*, 98.

a radically innovative contribution to the debate: he seems to have been the first to read not only sea-level change (as suggested by Tylor) but also glaciation and thawing as ongoing processes. He detailed the possible impacts both of a new ice age and of extreme warming, stressing the potential long-term effect on present ice caps of even a small change, in either direction, in the average global temperature. Whittlesey estimated that during the last ice age up to one-fifth of the globe might have been covered in ice. While the thickness of the ice was hard to assess, he guessed, based on geological data from New England, that it might have been as much as 5,300 feet (about 1,600 meters) thick. Whittlesey combined this estimate with similar observations from around the world and concurred with Maclaren that the growth of glaciers in the period would have drawn enough water from the seas to reduce their level by at least 350 feet (about 100 meters). Projecting the back-and-forth between ice ages and thaws into the future, he concluded that the melting of existing ice caps would cause a catastrophic rise: Greenland's, he estimated, would add 3 to 4 feet (about 1 meter) to sea level; Antarctica's could add 12 feet (about 3 meters).[11]

At around the same time, the Scottish geologist Thomas Jamieson proposed that a perceived change in sea level along the coasts of Scandinavia—the change that first drew attention to the theory that attributed all such changes to crustal movements—was also linked to the aftermath of the ice ages. Based on field observations of the Pleistocene glaciation in Scotland, he theorized that the crust was rising in response to the decreased weight of the ice cap on top of it. In the following decades, others—apparently independently, as no cross-references are given in early publications discussing the theory—proposed similar mechanisms. Thus was born, about thirty years after the idea of the ice age was first proposed, isostasy—the theory of postglacial uplift—and its connection to absolute sea-level changes roughly theorized.[12]

Jamieson had suggested, as a cause of the continental floods that seemed to have occurred during the ice ages, the depression of the earth's crust under the weight of an ice cap. In response, the Scottish geologist James Croll echoed Adhémar's views on the impact of the precession of the rotational axis—again, apparently without knowing of them, and taking a uniformitarian rather than catastrophist stance. In a letter published in 1865 in the short-lived London-based magazine *The Reader*, Croll suggested that the massing of ice on one hemisphere would shift the planet's center of gravity, leading to an increase in sea level in the same hemisphere. Croll also advanced the geologist James Murdoch Geikie's theory of multiple glaciations, promoting the idea that sea level had changed repeatedly through the ice ages, a view supported as well by the likes of Alfred Russell Wallace and Charles Darwin.[13]

In a letter published in the same journal a few weeks later, the English geologist and paleontologist Searles V. Wood countered Croll's interpretation, reiterating what Maclaren had noted two decades earlier: that glaciations would necessarily have reduced the amount of liquid water available on the planet and thus caused a decrease in sea level. Lyell made the same point in a private exchange of letters, which led Croll to revise his calculations. But, like Chambers, he remained unconvinced by the "old notion that the general level of the sea remains permanent." Given the local nature of crustal movements, he argued, only absolute sea-level changes could explain the dispersed but similar shifts in the relative position of sea and land suggested by ancient, raised beaches. These features were first brought up by Henry De la Beche, then director of the British Ordnance Geological Survey, in 1839, then discussed by Chambers and taken up again at an 1866 meeting of the British Association for the Advancement of Science by the geologist William Pengelly.[14]

At Croll's request, William Thomson, later Lord Kelvin, weighed in with an explanatory note on the influence of ice caps on the level of the sea that supported Croll's overall idea of glaciations alternating between the hemispheres. More than a decade later, in an 1888 presentation at the Geological Society of Glasgow, he further developed Croll's theorization of the gravitational attraction exercised by the Antarctic ice cap, endorsing the idea that sea levels would rise in the hemisphere subject to glaciation: "The ice attracts the fluid it displaces; but the fluid displaced itself attracts the remaining fluid and so contributes to the resultant attractive force." This phenomenon, Thomson believed, might prove to be the "most probable explanation of some of our familiar changes of sea level." While he contested Croll on, for instance, the thickness of the Antarctic ice cap, middle ground could be easily found; a thinner ice cap would, he thought, suffice to explain the long-term sea-level changes Croll hypothesized.[15]

An increasing number of scientists began, in the second half of the nineteenth century, to question the presumption that the sea was stationary, on either a geological or a historical timescale. They propounded multiple theories, positing various causes for changes in sea level: glaciation, the accumulation of alluvium, crustal movements. Some held that glaciation caused sea levels to rise, some that the ice ages brought about a fall. Each of these theories struggled to prove itself against the dominant Lyellian model of a stable sea facing moving land. Things were about to change, however, as new theoretical approaches sought to include sea-level changes within Lyell's theoretical framework, instead of opposing it.

EVERYTHING MOVES

In the mid-nineteenth century, supporters of changes in absolute sea-level remained critical of the dominant theory of a moving crust. This put them at the fringes of the scientific community. During the last third of the century, though, a more balanced approach became increasingly common. In 1869, the geologist Hermann Trautschold, carrying out accreditation work in Russia after obtaining his doctorate in Germany, presented a master's thesis at the university in Dorpat (now Tartu in Estonia), in which he criticized the widespread preference for vertical shifts of the earth's crust in explaining changes in the relative levels of land and sea. The most common ideas, he held, were not the right ones. The history of geology, he suggested, was itself cyclical: the eighteenth-century preference for the sea as the primary agent of land formation had yielded at the turn of the century to new theories favoring vulcanism and crustal movements. He expected the latter to, in turn, give way to a revival of theories highlighting the role of absolute changes in sea level in land-formation processes. A decade later—prompted by an article in the French cultural and literary monthly *Revue des deux mondes* claiming that sea level had been fixed since the beginning of the ongoing geological epoch—Trautschold published an essay discussing the intrinsic instability of sea levels. He deemed it unrealistic to suppose an absolutely stable sea when the local shifts of the earth's crust cited by geologists to explain changes in the relative position of land and sea would themselves alter the shape of the oceanic basins and consequently the height of the water they contained.[16]

Nathaniel Shaler, a notorious racist and apologist for slavery, was a disciple of Agassiz in Harvard. He became professor of paleontology and geology at the same institution—adroitly balancing fidelity to his mentor's heritage and the rise of Darwinism—as well as the director of the Kentucky Geological Survey. In an 1875 paper, he asserted that it was unnecessary to take sides: both movement of the earth's crust and variation in sea level should be considered when explaining relative sea-level change. "The ancient view of the sea being the mobile region was not without its truth, and was too quickly abandoned," he wrote. Tabulating the many theories discussed in previous decades, he assessed what he thought were the advantages and limitations of each. In particular, he criticized Maclaren's idea that the main cause of relative changes in sea level had been the capture and storage of water by expanding ice caps. His main objection to the theory was that, with no clear knowledge yet available about the depth and extent of ice during the last glacial episode, its impact was difficult to assess. Shaler estimated a possible decrease in sea level of 1,200 feet (about 360 meters) due to glaciation. This would itself have led, he claimed, to a major fall in temperatures, comparable to that experienced by raising any point of the earth by the same height. The lack of paleontological evidence that animals moved from the north to the tropics led him to distrust Maclaren's theory. Shaler's stance was a uniformitarian one: oceans and land, he thought, coexist in an overall state of equilibrium; the effects even of major disturbances, such as glaciations, tend cumulatively toward the original condition. The continental shelf, he suggested, may represent "the region of constant alternation between land and sea conditions"; thus, the moment he was living in was one of extremely high waters.[17]

Theorists' strict separation between sea-level rise and continental subsidence as the cause of changes in the relative positions of land and sea was indeed eroding. In 1882, the German geographer Albrecht Penck proposed integrating Buch's and Playfair's preference for crustal movement with the many theories favoring absolute changes in sea level. Buch's and Playfair's theory was grounded, he claimed, in assumptions about the mechanics of the planetary crust and core that had since been disproven. Yet its truth was rarely questioned. Early doubts and criticisms had been either ignored or retracted by their authors, notably those of Lyell, who had then become one of the theory's most committed advocates. Penck favored a reinterpretation in which both land and sea move to some extent. In not allowing for that possibility—fixating on just one causal mechanism—geologists had, over the course of the nineteenth century, found themselves concocting incredibly complex theories to accommodate exceptions to what was seen as the rule.[18] While vertical land shifts work well to explain discrepant trends in sea-level variations registered at different locations, absolute sea-level changes are a simpler answer when similar variations seem to have occurred at the same time in distant places. The explanatory framework Penck proposed would no longer treat the two mechanisms as mutually exclusive.

Penck claimed also that Chambers's assertion that changes in sea level would affect in exactly the same way places across the oceans, even at great distances, had been refuted by the work of the German mathematician Johann Benedict Listing. Penck explicitly connected the work of scholars interested in the long-term history of the earth's changes with that of geodesists looking at the current shape of the planet as ascertained through gravitation. Listing had shown how the geoid—the shape of the planet's gravitational field—differs from the ellipsoid—the geometrical abstraction of the planet's shape. This affects the distribution of seawater, which naturally follows the gravitational field rather than the ellipsoid. Due to gravitational attraction, then, sea levels are substantially higher close to the continents' coasts. Not all changes in the level of the sea, that is, are felt in the same way at all points on its surface, which consequently is not so ideal a reference level as had been assumed. Such differences between the measured sea level and what he deemed to be its theoretically ideal state had been noted a few years earlier by Julius von Hann, professor of physical geography at the University of Vienna. In an article discussing the gravitational variations of sea level and the significant discrepancies between the geoid and the ellipsoid (or spheroid), Hann noted how all altitudes actually refer to an "irregular, disturbed, and even varying level," rather than to the "surface of the spheroid to which the real sea surface belongs."[19]

The intrinsic disturbances and variations Hann noted in the level of the sea combine with regular tidal fluctuations and the impact of waves to make the relation between land and sea unstable. The idea of an unstable sea began to spread among surveyors as well. Helmert wrote about how the "constant agitation" of the oceans made it difficult to determine boundaries between land and sea. In a talk given at a meeting of the Internationale Erdmessung in 1887, Lallemand characterized mean sea level as a transient datum, in that the ocean's level is subject to such long-term influences as the planet's cooling and the impact of meteors.[20]

In 1888—more than a quarter century after his strolls around Eggenburg suggested insights about the relation of absolute sea level to land formation—Eduard Suess published the second volume of his magnum opus *Das Antlitz der Erde* (*The Face of the Earth*). In it, he detailed his theory of how traces of ancient shorelines formed through shifts in the level of the sea. His causal explanation did not, however, entail the sequestration of water on land during the ice ages. Suess wondered, rather, if crustal movements themselves might be responsible for the major changes in sea level detected in the geological record. What would happen, for instance, if a huge sinkhole opened below the Pacific Ocean? Water would fill it, retreating swiftly from the coasts of the world, leaving behind the markers of ancient beaches. Continuing the cycle, continental debris, carried into the oceans as silt, would slowly fill the hole and thus cause a renewed global sea-level rise. Eustatic movements of the sea, as he termed them, appeared more apt to explain gradual and uniform transformations at the global level than did the direct impact of crustal shifts, which he deemed better suited to explaining local, circumscribed, possibly catastrophic modifications.[21]

Suess noted, following in Darwin's and Chambers's footsteps, how difficult it is to find geological proof of sea-level rise: the rise of water by its nature hides its own traces. The only reliable marker may be provided by coral reefs, which grow gradually in parallel to the rise of the sea, leaving behind huge amounts of dead but durable coral. The fall of the sea, by contrast, is easily registered. Suess's concluding criticism was directed to the Lyellian idea that crustal movements occur in regular alternation—and were still happening as measurable occurrences at the time he was writing. Suess dismissed all the changes he was able to account for in historical times as minimal and local: what are a few thousand years in respect to the timescales of geology and astronomy, he asked rhetorically. Suess's focus on the idea that significant sea-level change occurred only on a geological scale paradoxically allowed him to continue to support the idea that for all practical purposes the sea ought to be considered stable through historical times. Contrary to many other attempts at bringing absolute sea-level variations back into geological theory, such as Tylor's almost forty years earlier, Suess offered a framework that preserved the ongoing work of geodesists and surveyors, while enriching the explanatory models of land formation. That sea-level changes happen on the geological timescale, so slowly as to be essentially imperceptible, allowed sea level to function as a practical vertical benchmark, while still acknowledging its long-term role in shaping the land.[22]

NEW STANDARDS

In the latter decades of the nineteenth century, scientists and practitioners increasingly accepted that sea level is as arbitrary a choice for a vertical datum as any. There was, however, also a degree of consensus that the surface of the planetary sea represented a good approximation of the outer limit of the earth's geometrical shape, as suggested by Bessel already in the 1830s. As discussed in chapter 3, mean sea level had become the de facto international reference point for elevations. But the issue of whether sea-level changes were perceptible only on a geological timescale or also a historical one, if not a day-to-day basis—discussed so thoroughly by geologists in the previous decades—was now a matter of primary interest for geodesists and surveyors as well. On the one hand, measured mean sea level—though subject to errors and developments in the accuracy of instruments—could be adopted as a stable, if approximate, representation of the shape of the earth. On the other, datums based on sea levels would have necessarily to be continually reassessed. This developed, in the early decades of the twentieth century, into further research on how to determine a vertical reference point based on sea level and how long a data series should be compiled for this purpose.

Work done to explain short-term variations in sea level played a crucial role in clarifying that single measurements and short series were insufficient to produce stable and reliable datums. In 1915, the scientific adviser of the Fisheries Board of Scotland, the biologist D'Arcy Wentworth Thompson, lamented, for instance, that the causes of year-to-year sea-level variance had not been sufficiently explored. Working with about fifty years of almost-continuous tide data gathered in the ports of Aberdeen and Dundee, Thompson ventured to assess trends in yearly and monthly variations. His first insight was that mean sea level follows a curve throughout the year, reaching its lowest point around May and its highest in December. This phenomenon, he reported, had been of great interest to many along the coasts of the Baltic and North Seas since William Thomson, Lord Kelvin, had first noted it in the mid-nineteenth century: the curves at various stations in the region were analyzed and found to have pretty consistent patterns. On the other side of the Atlantic, the Canadian superintendent of tidal surveys reported in 1917 an annual variation of mean sea levels in both Quebec and the state of New York. It was becoming clear that just one or a few years' gaugings were insufficient to determine with any precision the position of the ocean's surface.[23]

While a thorough debate over how to select an international point zero had occupied scientists and practitioners in Europe for decades, at the turn of the century little work had yet been done in North America to define a standard vertical datum. Despite its many limitations, mean sea level was clearly becoming a global standard of sorts. In Canada, before the government-sponsored tidal survey led by William Bell Dawson began in 1893 to record tidal fluctuations at ports on the Atlantic and Pacific coasts, no standard reference plane had been defined. It was only as a collateral effect of the collection of data on the temporal dimension of tides that the first coherent series of sea-level measurements were gathered. As Dawson wrote in 1903:

> Eventually as the observations are continued the value of Mean Sea level, extreme tide levels, and other factors of importance, are determined with reference to the local benchmark. Although there is as yet no general system of levels in Canada, these results are of service locally in the meantime; and they also furnish a basis for any more extended geodetic levelling which may be undertaken.[24]

In the following years, agencies in charge of public works in different parts of the country independently adopted mean sea level as ascertained by the Tidal Survey as their vertical datum, marking the beginning of a slow but steady standardization process.[25]

In the United States, the Coast and Geodetic Survey—undertaking the country's first transcontinental survey, from Chesapeake Bay to San Francisco along, approximately, the 39th parallel north—referred its leveling operations, concluded in 1896, to mean sea level as measured at multiple tidal stations, from Boston to Seattle by way of Galveston, Texas. Sea level was not yet, however, an accepted national standard. Each county, municipality, and railway company adopted its own peculiar vertical benchmark. These could refer, as in France half a century earlier, to virtually anything, from a nearby lake to an arbitrary benchmark. In New York City alone, consequent to the consolidation of multiple boroughs into one city in 1898, more than ten different reference planes were still in use for day-to-day operations in 1915. Engineering errors were a constant risk and bloated costs almost a certainty. According to a 1916 survey of American municipal engineers, just five cities in the United States had adopted mean sea level as an official datum. While many respondents agreed on the desirability of a common reference point and most states already used mean sea level when practicable, there was apparently no political will to enforce a standard that would cross municipal boundaries or company limits.[26] The superintendent of the Coast and Geodetic Survey, E. Lester Jones, was nonetheless confident that locally measured mean sea level—previously adopted by his agency—would soon become the most commonly accepted reference point for elevation and the de facto national standard:

> Mean sea level is the only datum that is a universal one; that is, it is the only datum that may be used for many detached pieces of leveling at different parts of our coasts. . . . There is no limit to the places along the coast at which mean sea level may be determined and be considered as having zero elevations for any line of levels starting inland from them.[27]

Another decade, however, would pass before the first US standard vertical datum was defined formally. The so-called Sea Level Datum of 1929 (SLD29) was based on the local mean sea levels of twenty-one US and five Canadian stations. Although it was well known that each tide gauge might record a different height for the sea, in defining the datum surveyors assumed that all of the stations were at the same level. The recorded discrepancies were in fact considered to be within the margin of error for measurements spanning the whole continent. Moreover, users expected heights to be compatible with tidal observations and might be confused by negative values. Canadian authorities had, however, just published their own adjustment, based only on the five Canadian stations, the year before. Out of fear that imposing a new change so soon might also confuse users, they decided not to adopt immediately the one produced in cooperation with the United States.[28]

Back in Europe, the British Royal Commission on Coast Erosion and Afforestation addressed in the 1910s the perceived oscillations in relative sea level around Great Britain. The whole of western Scotland, for instance, is submerged: the lochs are actually drowned valleys, and according to the commission, their inundation occurred in relatively recent geological times. Clement Reid, who worked with the Geological Survey of Great Britain, highlighted, in testimony to the commission, the extent to which the details of and reasons behind the changes in relative sea level in historical times remained unknown. He remarked that future variations, even of just a few feet, could have "disastrous effects." Whether the perceived stability of sea levels would endure or was just an interlude, "which may at any time give place to rapid change," seemed unknowable at the time. Reid, who played a central role in early research on the land connecting Great Britain to the European mainland during the last ice age, also criticized the commonly accepted idea that so-called submerged forests were due to a combination of local causes; he stressed, instead, how they could only have been produced by changes in sea level, due either to subsidence or to absolute rises.[29]

At the beginning of the twentieth century, long-term discrepancies between mean tide and mean sea levels and the differences in measured sea level at different British ports prompted the Ordnance Survey to abandon the Victoria Dock datum of 1844. The idea was to define a new datum, the average of stations located at the three corners of Great Britain, that was supposed to account for the whole extent of sea-level variations along British coasts. The stations were to be set up in Newlyn in Cornwall, Felixstowe in Suffolk, and Dunbar by Edinburgh. Once the project was under way, however, surveyors realized that the difference in level between Newlyn and Dunbar was greater than the standard error for level measurements. In the end only Newlyn station was adopted. The new effort did not always confirm the perceived discrepancies that had prompted it. Surveyors concluded, for instance, that, considering the possible leveling errors, "there is no indication that there has been any change in the relative levels of the coast lines" between Bristol Channel and the English Channel, as noted by the American physical geographer Douglas Johnson, a supporter of the idea of a sea that has remained substantially stable over millions of years. The huge endeavor—buttressed by the idea that by averaging sea levels it would be possible to determine a "natural" reference framework for elevations—seemed to strengthen once more the idea that datums are quintessentially conventions. Whether or not sea levels had actually changed in historical times began to appear inconsequential to the work of the geodesists. What mattered was the long-term reliability of the chosen vertical benchmark, independent of whether it was grounded in any natural feature. Hunting a moving reference point with continuous surveys and reassessments risked producing more damage and confusion than sticking to one that did not faithfully reflect changes in the level of the sea.[30]

GLACIAL CONTROL

The (relative) invariability of shorelines in historical times remained a common assumption in the geological community until well into the twentieth century, as reported in a detailed review article published in 1920 by Franz Suess, Eduard Suess's son. In 1912, Douglas Johnson had, for instance, criticized earlier writings supporting the theory that the Atlantic coast of North America was still measurably subsiding. His skepticism was directed against all claims of significant changes in the level of the sea. The word "significant" matters here: Johnson in fact inferred that changes of up to 30 centimeters might have been possible throughout the last millennium, but he assumed they were probably due to local changes in the height of high tides. Apparently his idea of significance was on the order of meters.[31]

Reviving a suggestion made by Alfred Tylor decades earlier, the Canadian geologist Reginald Daly worked on strengthening the connection between sea-level changes and glaciation. After an expedition to Hawaii gave him an opportunity to study coral reefs, Daly published a series of papers in which he developed the idea that reef formation had been geologically controlled by glaciation. Not only had glaciation moved southward the northern limit at which coral reefs could form, he wrote; it had also caused, by way of variations in the glacial accumulation of water, subsequent increases and decreases in sea level. While earlier studying the postglacial elevated strands of Labrador, Newfoundland, and Quebec, Daly had begun to realize that the level of the sea might have decreased relatively recently, in geological terms. While he recognized that testing Suess's eustatic theory would be difficult, he held that it deserved further discussion. Clues hinting at eustasy could, in fact, be found all over the planet, from Australia to the West Indies, Patagonia to Great Britain. In particular, the fact that similar changes had been registered for both the eastern and western coasts of Australia suggested that they could not have been caused by a uniform emergence of land. As part of his glacial control theory, Daly proposed in 1920 that the storage of water as ice on land might have caused a generalized sinking of sea level of approximately 20 feet (about 6 meters) during what he termed the "human period." Daly's theory would become in the following decades the most widely accepted theory explaining global mean sea-level variations since the end of the last glaciation.[32]

As we have seen, even if the theoretical connection between sea levels and glaciation had been in the making for almost a century, in the 1920s the question of what caused perceived changes in the relative positions of land and sea remained open. Was it crustal movements or absolute changes in the level of the sea? In the 1930s, however, Daly's glacial control theory gained momentum. In 1930, a US Geological Survey scientist, Charles Wythe Cooke, published an article on the ancient shores of the Atlantic, in particular, the horizontal terraces along the coasts of Georgia. The uniformity of the terraces convinced him, as it had others before him, that only sequential sea-level rises and falls could fully explain them. The most plausible cause appeared to him to be the mechanism described in Daly's theory. As Whittlesey had in his much earlier work, Cooke also estimated the possible impact of the complete melting of the ice caps of Greenland and Antarctica, which he assessed at 200 feet (about 60 meters).[33]

In 1933, Albrecht Penck noted how Antarctica and Greenland are physical remnants of how the ice cover of the ice ages looked. Their presence is a clear sign of how human history is located in an interglacial period. If we were further on in the deglaciation process, sea level would in fact be much higher. Were the remaining ice caps of Antarctica and Greenland to melt, Penck predicted a further rise of 55 meters. Penck's work makes clear how the theory of combined isostatic and eustatic movements—in particular, the elevation of landmasses freed from ice caps concurrent with a global increase in sea level through the addition of melted ice—had been accepted by the scientific community as the primary explanation of relative sea-level changes in both geological and historical times.[34]

The renowned American geographer William Morris Davis presented, in early 1933, a paper at the California Institute of Technology in which he discussed ancient beach deposits in the Santa Monica mountains near Los Angeles. Believing he had determined a pattern in the traces left by subsequent ice ages, he concluded that we live at the beginning of a long interglacial period and that it would be some time before the planet cooled again. Press reports about these claims did not really consider the possible negative consequences of rising sea levels caused by the further melting of polar ice caps, but did herald the hope that "the poles may become useful and inhabited places."[35]

In November 1933 Henri Baulig, a French geographer who had studied with Davis, gave a series of lectures at the University of London on the issue of sea-level change. While he began by reiterating the substantial invariability of water levels along the world's coasts for any practical purpose, he recognized the massive changes that had clearly occurred on longer, geological timescales. Most geologists and geographers, he claimed, still had a Lyellian preference for attributing perceived changes exclusively to movements of the land: "Raising or depressing a limited portion of the land appears easier and of less consequence than moving the huge mass of the ocean." Vertical crustal shifts can be quite easily explained on extremely local scales, while variations in sea level require a global cause and are easily disputed by the lack of consistent evidence along all coasts. Building on recent work that claimed the end of the last ice age had been accompanied, along the coasts of Scandinavia and northern France, by a gradual rise of the sea, Baulig provided a stronger theoretical background to Maclaren's early insights about the impact of glaciations on sea level. He cited in this context the work of the Swedish-American geologist Ernst Antevs, who had estimated that during the last glaciation sea levels were at least 100 meters lower. Thus, glacio-eustatism, the theory explaining repeated changes in sea level through the ice ages, was officially born. The widespread acceptance of a clear connection between the level of the sea and the ice ages made it possible, once again, to hypothesize sea-level variations on a temporal scale shorter than the geological.[36]

Since Agassiz first proposed the idea of ice ages, the idea that glaciation, and in general the accumulation of ice in polar caps, played a central role in determining sea level in both geological and historical times has cyclically reappeared in the scientific discourse. The difficulty of performing truly global scientific research and gathering, assessing, and digesting data, beyond the anecdotal, from the most disparate parts of the globe long limited the ability of proponents of theories based on absolute sea-level change to make their voices heard. At the same time, the growing number of geologically recent instances in which changes in the level of the sea had been registered forced supporters of crustal-movement explanations to come up with more and more complex theories about repeated, synchronous shifts of the planet's crust across great distances.[37] A stable sea was becoming increasingly hard to prove, but so was a uniformly moving terrestrial crust. As a consequence of the onset of anthropogenic changes on a geological scale after the industrial revolution, the first consistent data series showing an ongoing rise in sea level were also being registered. What has been termed a no-analogue future—that is, a geological future that, contrary to the precepts of uniformitarianism, does not resemble the past or replicate its processes—is already the present. The ways in which this cataclysmic shift has begun to be registered in changes of sea level, and in the instruments developed to make such measurements, will be the objects of analysis of the next chapter.

FIVE

GOING GLOBAL

In 1933, the Finnish oceanographer Rolf Witting, secretary of the International Association of Physical Oceanography (IAPO), submitted a memorandum on "mean sea level and its changes" to the association's Lisbon meeting. The memorandum did not come out of the blue; it was the result of over a decade of institutional work. Throughout the 1920s, the tidal committee of what was then still the Oceanographic Section of the newly established International Union of Geodesy and Geophysics (IUGG) had discussed the need to gather data about tides and sea level in cooperation with other IUGG sections. The IUGG had taken over and broadened the activities of the many scientific associations whose efforts to promote "governmental internationalism" in the sciences had been interrupted by World War I.

Witting stressed the importance of the study of mean sea level "for the solution of a complex of geophysical problems," including the relation between the amount of water present in the oceans and the variations in polar ice cover. He suggested that the IAPO join forces with the International Association of Geodesy (IAG) and institute a permanent committee tasked with improving the global coverage of tidal data. Witting lamented, in particular, the lack of permanent stations over vast stretches of the planet: "The filling of these gaps would be of great value." He envisioned a network of stations with very precise minimal requirements, each located no more than 1,200 nautical miles from its nearest neighbor. Twenty new coastal stations, he reckoned, would suffice to fill the gaps, while ten stations located on islands would be necessary to connect the continental coasts. By 1936, the International Committee for Mean Sea Level and Its Variations had been set up in Bidston, near Liverpool, under Witting's chairmanship and in cooperation with the IAG and the International Hydrographic Bureau. In 1939, it published a first collection of sea-level data, gathered from stations all over the world for the years prior to 1936, along with information about the recording institutions. This first publication was followed by updates in 1950 and 1953. This committee is commonly considered the precursor of the Permanent Service for Mean Sea Level (PSMSL), which is still responsible for collecting and sharing sea-level data.[1]

This new institutional setting, in conjunction with the establishment of glacio-eustatism as a theoretical framework, contributed in the interwar years to an environment open to work on present-day sea-level changes. The relation of ongoing deglaciation processes to measurable rises in sea level gradually became central to scientific discourse about global climate, its acceptance not limited to scholars working primarily on glaciers or ice ages. Forecasts of rising seas grew ever more common. In the postwar years, the need for a more global picture of sea-level variations gained momentum, as did the consensus that sea-level changes are a present concern, not just a matter of the geological past. This occurred in parallel to a renewed interest among earth scientists in global phenomena and an increase in funding for large-scale comparative projects.[2]

The idea of a rising sea began also to be established as a cultural trope, primarily through the pages of popular science magazines but also in science fiction and even in songs. Meanwhile, the development of satellites offered scientists and geodesists a new point of view, an opportunity to gather proxy data about the shape of the planet and its climatic conditions free from the limitations of shore-based measurements. The widespread hope was that new technical means would make it possible to achieve both precision and objectivity. In the following decades, technological improvements and the consequent amassing of observations allowed researchers to formalize the idea of ongoing "global warming" and to start discussing its potential consequences for sea levels. By the early 1980s, data about long-term changes in the level of the sea were repurposed as indicators of global climate change, and a political debate about how to face such change began to develop. In 1987, anthropogenic sea-level rise became a matter of debate for the US Congress, and in 1990 an object of inquiry of the Intergovernmental Panel on Climate Change (IPCC). It is in this context of sustained scientific and political debate, accompanied by steady technical developments, that sea level has become truly global.

EXPLAINING PRESENT CHANGES

Oceanographers' work in the 1930s was not limited to setting up associations and committees, or signing memoranda of understanding with those instituted by other scientists. By the end of the decade their data collection efforts had helped consolidate the idea that ongoing deglaciation and possible sea-level changes were causally related. In 1938, for instance, at the concluding round table of a symposium on the geophysics of ocean bottoms held at the American Philosophical Society in Philadelphia, the American atmospheric physicist William J. Humphreys posed the question of how melting ice affected global sea levels and how that effect would evolve in the future. If glaciers continued to melt at a steady rate, he expected that the planet would face, within a few thousand years, a radical shift in climatic conditions. As the ice cover receded, drier areas and storms would, for example, move toward the poles. On a similar note, building on the increasingly rich literature produced since the 1920s about Scandinavia's postglacial rebound and on an ongoing debate about Arctic climate change, the Icelandic geologist Sigurdur Thorarinsson vehemently argued in 1940 for bringing accounts of historical and geological changes in sea level up to the present time.[3]

Following up on work he had done on the Vatnajökull system of glaciers in Iceland in the late 1930s, together with Hans Ahlmann, a Swedish theorizer of polar warming, Thorarinsson showed how, since 1890, the glaciers had been steadily losing mass. The thinning of those glaciers alone could, he calculated, have caused a global rise in the level of the sea of 0.14 millimeters. In an article published in the Swedish *Geografiska Annaler* in 1940, he went beyond that individual case to develop a theory of the global impact of glacier shrinkage on sea levels. In a thorough review of the previous two decades' glaciological literature, he reconstructed the trends of glaciers all over the world, concluding that ongoing reductions in ice cover were producing an annual sea-level rise of approximately half a millimeter per year.[4]

Thorarinsson admitted that this was an extremely rough approximation, based on scant data and multiple assumptions—for instance, that marginal areas accounted for most of the thinning, and thus the huge ice masses toward the interior of Antarctica and Greenland could be omitted from the calculation. But despite the inconsistent quality of the source data, which depended hugely on where and by whom they had been collected, Thorarinsson concluded that "the present glacier shrinkage is a universal phenomenon." The last glacial peak had occurred in the mid-nineteenth century. By focusing on the subsequent gradual reduction in the overall volume of earth's glaciers, Thorarinsson provided the first clear connection between measured changes in glaciers and estimated eustatic changes in sea level. Previous studies of melting glaciers' potential impact on sea level—such as those by Whittlesey and Cooke—had considered hypotheticals, not actual rates of change. Humphreys had given a rate but no detail about how he'd arrived at it. Thorarinsson believed that his own estimate, while only an approximation, offered an idea of the minimum magnitude of the phenomenon. He mentioned as well some further work by Ahlmann, who in 1940 reported a 10 percent reduction in volume in most glacier districts since their peak. Applying this number to an estimate of the volume of all of the world's glaciers outside of Antarctica and Greenland proposed by the Finnish geologist Wilhelm Ramsay, Thorarinsson derived a sea-level rise of almost 10 centimeters in a bit less than a century.[5]

Other scholars approached the analysis of sea-level variations in other ways. Beno Gutenberg, a German geophysicist who, lured by the offer of a professorship at CalTech, had moved to the United States in 1930, was primarily interested not in glaciers but in crustal movements, in particular seismology. His attention to absolute variations in sea level grew from the need to better understand their interaction with postglacial uplifts. It was thus as part of a work on the reciprocal influences of eustasy and isostasy that in 1941 he published an estimate of 10 centimeters sea-level rise over the previous century, almost identical to the rate suggested by Thorarinsson. But where Thorarinsson had derived sea-level changes from the rate of glacial loss, Gutenberg looked directly at recorded sea levels—the tide-gauge data gathered and collated by the IAPO in the previous decades. And while he did calculate a mean trend of global eustatic change for the whole previous century—compensating for possible errors, meteorological influences, and land movements—Gutenberg noticed irregularities in the spatial and temporal distribution of changes: "the periods of faster rising did not occur simultaneously at all stations," and the rate appeared to have accelerated in recent decades. This acceleration could lead shorter-time series, like the one from Britain's Newlyn station, to overestimate longer-term trends. Or they might instead—possibly because local conditions lag behind oceanic fluctuations—run counter to the prevailing, global trend, with sea level appearing to fall swiftly rather than to rise. The inconsistency of these observations persuaded Gutenberg to give preference in his computations to longer series. A significant assessment of trends and their variance over time could be produced only on the basis of long data series, he claimed.[6]

The connection between sea-level rise and deglaciation highlighted by Gutenberg was confirmed in 1945 by the Dutch geologist Philip Kuenen. A specialist in marine geology, Kuenen found a neat correlation between the historical trend in sea levels along the coasts of the Netherlands and available data on glacial recession. The estimate he offered of expected long-term rise was far higher than those of Thorarinsson and Ahlman; even a minimal melting of the Antarctic and Greenland ice caps, he concluded, could contribute significantly to sea-level rise. As soon as "a trustworthy theory on climate variations" was developed, he suggested, it would become possible to predict such changes.[7]

Taking into account recent developments in geophysical research, Gutenberg noted several possible causes for the recorded global rise in sea level. These ranged from the uplift of ocean bottoms, to sedimentation, to the melting of glaciers. The fact that the trend coincided almost perfectly with Thorarinsson's data might point—nearly a century after Maclaren's intuition—toward a realization of glacial loss as *the* driving cause. By the early 1940s, the idea of historical and ongoing sea-level changes had reached well beyond geology's strict disciplinary boundaries. Indeed, they are mentioned, along with human activities, in the historian Marc Bloch's enumeration of possible causal factors behind coastal transformations in historical times. In the first chapter of his renowned methodological oeuvre *The Historian's Craft*—written without access to notes and sources due to the German occupation of France, and first published posthumously in 1949—Bloch talks about the historical agency of sea-level variations. Specifically, he discusses the complex array of causes that might have first produced the Zwin inlet—which gave

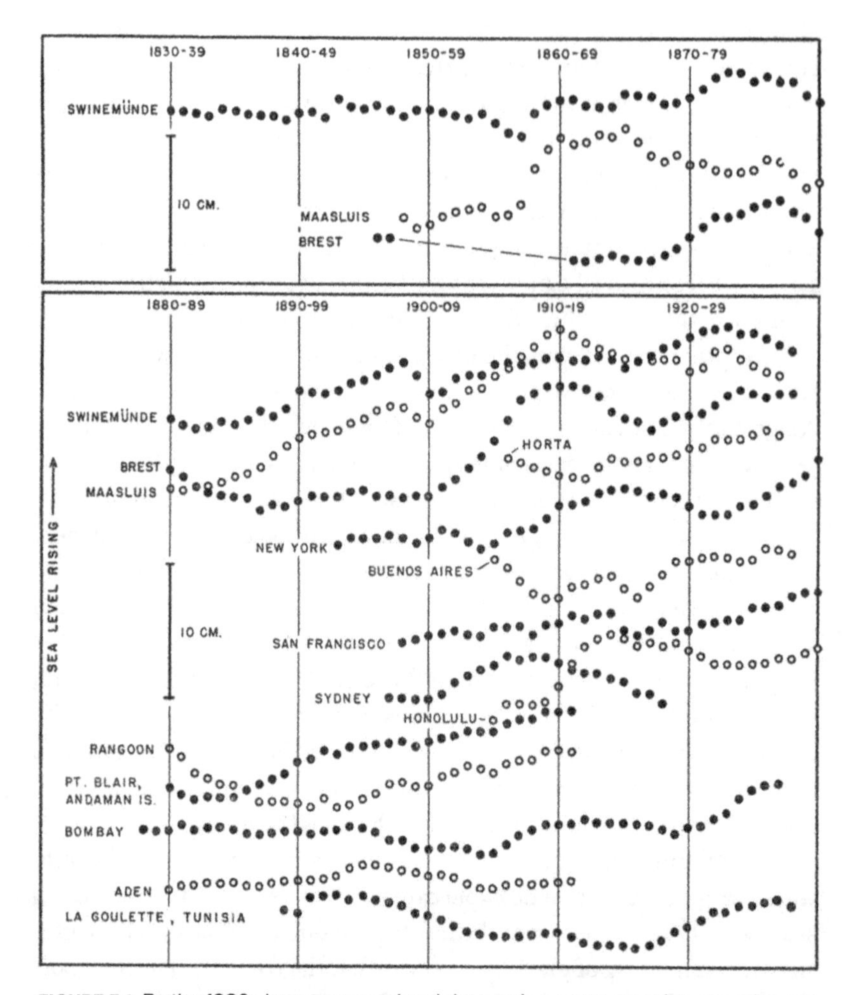

FIGURE 5.1. By the 1930s long-term sea-level data series, some covering an entire century, were available for comparison, which seemed to suggest a trend toward a global rise of sea levels. The image tracks sea-level variation recorded at tide gauges around the world between 1830 and 1929, as compiled in Gutenberg, "Changes in Sea Level," 730. Courtesy of the Geological Society of America.

access to the sea, and consequent riches, to the city of Bruges—and the inlet's later siltation and the city's decline. Bloch affirms that neither the geologist nor the historian alone can satisfactorily answer the many interconnected questions a case like this raises as regards the interaction of humans and the sea along the coasts.[8]

RISING CONSENSUS

After World War II the scientific consensus around Thorarinsson's and Gutenberg's insights grew steadily. So too did the need for a global picture that could help explain which changes in the relative position of land and sea were local and which due to absolute variations in sea level. In 1948, Harry Marmer, the assistant chief of the Division of Tides and Currents of the US Coast and Geodetic Survey, directed his attention to what by then was a classic text: Douglas Johnson's 1917 paper on whether the Atlantic coast of the United States was sinking. Building on tidal observations, which had grown more numerous and available in the intervening thirty years, Marmer documented a steady rise in sea level throughout the seaboard. He remarked, however, that the trend did not indicate whether the sea had risen or the land had subsided: "To determine which is the active agent and which the passive is another problem." A clear answer, he stated, was impossible on the basis of such limited regional data. Personally, he was inclined to agree with Johnson and see the variation as a "lowering of the coast relative to sea level." Nonetheless, he maintained, only "systematic tide observations throughout the world will permit a precise apportionment" between absolute subsidence and sea-level rise. If the trend, as suggested by Gutenberg, was found to be global, changes would have to be due to fluctuations in the amount of water present in the oceans. If not, recorded variations could be attributed to local, isostatic causes.[9]

A couple of years later, the IAPO's mean sea level committee decided to collect and offer to the public raw data and extrapolations of long-term trends in mean sea-level variation, based on tide-gauge observations published over the previous decade. Sea-level change was by now considered, by oceanographers and geodesists alike, a "matter of great interest in view of the recession of ice and the increase in the water content of the oceans," as the committee's secretary, the British oceanographer Arthur T. Doodson, put it. The data made available for three global regions—Europe and North Africa, the Americas and the North Pacific, and Great Britain and the Commonwealth—were raw, uncorrected for eustatic variations or meteorological influences. Consistent with Gutenberg's results, they seemed, however, to indicate a clear trend toward absolute global sea-level rise, except for regions experiencing postglacial uplift.[10]

Around 1950 a new scientific consensus—that sea-level change was very much an ongoing process—seemed to be coalescing. Yet the dogma persisted that mean sea level could, for any practical purpose, be considered stable. "What is so sacrosanct about sea level?" the American geologist Kenneth K. Landes wondered publicly at the 1952 meeting of the Geological Society of America. The Australian geologist Rhodes W. Fairbridge—who would gain recognition for the so-called Fairbridge curve, plotting sea-level variations over ten thousand years—offered this perspective, reflecting his eustatic theories: "through geological eyes we might regard the present M.S.L. as ephemeral as a fleeting ray of sunshine on a wintery afternoon." In pursuit of such a comprehensive view, the IAPO had endeavored to coordinate the multiplicity of national and individual efforts that had flourished since the end of World War I. A global outlook was essential to distinguishing absolute sea-level rise and subsidence at local scales, noted the Danish meteorologist J. Egedal in his introduction to the report on long-term mean sea-level variations in Europe and North Africa, as had Marmer in his 1948 paper. Both affirmed the need for more data to further the research undertaken by IAPO.[11]

In 1952, Walter Munk and Roger Revelle, both at the Scripps Institution of Oceanography in San Diego and, later, crucial in developing the study of anthropogenic climate change, discussed the potential effect of rising sea level on the planet's rotational speed. The global rise of about 10 centimeters per century, signaled by Gutenberg and recorded in the data collected by the IAPO mean sea level committee, was consistent, Munk and Revelle noted, with observed changes in the planet's rate of rotation. But the rate of sea-level increase varied, they also noted, from decade to decade and from region to region, due to meteorological factors and crustal movements. This variability convinced them that it was best to avoid claims of global sea-level change. Even the most careful measurements and analysis, they argued, risked being drowned in the flood of data.

From an astronomical perspective, Munk and Revelle concluded, even eustatic changes a thousand times larger than those recorded in geological and historical times would not affect planetary rotation or the position of the poles to such a degree as to leave "a perceptible imprint on fossil records of life and climate." What would make their work, nonetheless, relevant to future research was the attention they called to the thermal expansion of water as a factor in sea-level change: an increase of a single degree Celsius in the temperature of the global ocean would, they calculated, lead to a rise of 60 centimeters, in addition to any increase in the amount of water present. Global sea-level rise was not, that is, directly proportional to deglaciation; it would happen much faster than earlier scholars had calculated.[12]

NEW SCENARIOS

By the 1960s, a narrative allowing for anthropogenic climate change and climate-induced sea-level rise was taking shape among scientists. This expedited the development of analytical means to interpret the rise of the seas as an ongoing process and attribute its recent acceleration to humans. In November 1959, the controversial nuclear physicist Edward Teller gave a talk on the need to find alternatives to oil as a dominant energy source, due to its role in what would later become known as climate change, at, of all places, the American Petroleum Institute.[13] In the Q&A he predicted that carbon dioxide-induced global warming could have catastrophic effects:

> But when the temperature does rise by a few degrees over the whole globe, there is a possibility that the icecaps will start melting and the level of the oceans will begin to rise. Well, I don't know whether they will cover the Empire State Building or not, but anyone can calculate it by looking at the map and noting that the icecaps over Greenland and over Antarctica are perhaps five thousand feet thick.[14]

The idea of a rising sea was, at the same time, making its way into popular culture. A planet upset by climate change and extreme sea rise was, in 1962, the setting for J. G. Ballard's novel *The Drowned World*, a contemporary take on the classic tale of the flood. Ballard pictured a tropical hellscape, characterized by burning heat, swamped metropolises, novel ecologies, climate refugees, and violent buccaneers. Rooted in a traditional view of global changes as "natural" phenomena—Ballard attributes the changes not to anthropogenic causes but to a sudden increase in solar storms—the book provides a prophetic view of how a "drowned world" might become a more violent one. Coming from a radically different perspective, both humorous and religiously tinged, the ukulele-playing comedy singer Herbert B. Khaury, known professionally as Tiny Tim, included in his first album, in 1968, a song addressing the consequences of sea-level rise: "Let everyone sing about those melting ice caps / How they're coming down into the sea / And let us all have a swimming time . . ." A direct reference to the biblical deluge ("All the world is drowning / To wash away the sin") also nods to the role of humans in causing the melting of the ice caps in the first place.[15]

Alongside the dystopian narratives, dreary forecasts, and silly songs, longer-term studies of Holocene sea variations became an increasingly common object of research. In the early 1960s, the Israeli Dead Sea expert Moshe Bloch registered, based on measurements taken on the Dutch coast, a rise in global mean sea level over the previous century of 15 centimeters. In agreement with Douglas Johnson's claim of about fifty years prior, he noted, though, that since exact data for any time prior to 1850 is unknown, recorded differences of less than 30 centimeters cannot really be considered significant. According to his estimates, "present sea level" was reached in the sixteenth century. Since then, he wrote, "there have been no indications of any permanent change of an order greater than 30 cm." Bloch worked on the assumption that most oscillations after 6000 BP were caused by changes in the ice cover of Antarctica, the mass of which was incommensurably greater than that of all the ice remaining in the northern hemisphere. He also envisioned these oscillations, even if relatively minor in proportion to the immediate postglacial rise, as having a perceivable impact on human history: the salt industry, for instance, depends on "low lying and flat ocean coasts."[16]

Since the mid-1950s there had been a steady increase in research dedicated to plotting the curve of postglacial changes in sea level. This became possible thanks to the development of reliable radiocarbon dating methods that allowed for novel reinterpretations of the old debate about submerged beaches. Hans Suess, Eduard Suess's grandson, and Francis Shepard, a disciple of Reginald Daly, compared, for instance, earlier research that carbon-dated shells and peat from the Dutch coast with new data, collected mostly along the coasts of Texas and Louisiana by the research lab of an oil company. Their comparative approach allowed them to rule out the notion that variations were due to local tectonic shifts. A truly global assessment, they noted once again, would be necessary to "distinguish conclusively between eustatism, tectonic movements, and compaction." With all its shortcomings, the International Geophysical Year of 1957–1958 played a crucial role in transforming the perception and assessment of ongoing sea-level changes by the international scientific community.[17]

Awareness of the issue reached a broader audience in the pages of popular science magazines. In 1960, Fairbridge published an article in *Scientific American* connecting myths of a global flood with the memory of a relatively swift postglacial rise in the level of the sea. He concluded that sea level is not only variable but "a sensitive indicator of even minor world climatic change." From this, he derived the idea that the geoid itself is subject to change as a consequence of climate cycles, oscillating between a minimum reached during peak glacial stages and a maximum in the warmer eras. Indications that throughout the postglacial age sea level and mean temperatures had risen almost constantly led him to conclude that the Quaternary ice age was reaching its end: "A slow swing to a warm climate could melt the last of the great glaciers. Then another great deluge will drown present shorelines and submerge many centers of whatever civilization is there to witness it." In 1963, the geologists William Donn and David Shaw moved beyond the deep history of glacial ages to correlate global temperatures recorded over the past century with sea-level trends along the Atlantic coast. Admittedly theirs was a limited case. But even if they lacked complete data, they anticipated that the sea-level trends were truly global, as were the temperature trends. The collection of observations from tide gauges on a planetary scale was, however, still deficient: political and economic issues hampered the development of science and international cooperation, while the history of how and when stations were set up still skewed the distribution of observations.[18]

Overcoming this deficit in the amount and distribution of data was crucial to computing global trends and comparing them through time. After World War II scientists and practitioners increasingly saw the necessity of liberating the measurement of sea level from its relation to the planet's landmasses, the quirks of coastal geography, and the spatial limitations imposed by the concept's colonial history. A way to reach this aim seemed to be offered by the swift development of satellite technology starting in the 1950s. Satellites promised to finally reach the "non-place" scientists had, throughout late modernity, construed for themselves as the ideal, neutral observation point. A material basis appeared at last to be available for the multiple layers of abstractions at the basis of all cartographic and geodetic efforts.[19]

DREAMS OF PRECISION

Space technology has radically changed the way we perceive our planet. Photographs taken from space have, for instance, tangibly represented the idea that the human species has only one world. The *Earthrise* picture, taken by members of the Apollo 8 mission in 1968, is iconic in this regard.

The change in perspective primed by the space race has affected also our understanding of what sea level is, as remote sensing adds further layers of meaning. It is not only a matter of scaling up, moving our point of view from the local to the planetary, from the immediate to the mediate, from the biased to the ostensibly neutral. It is also an issue of how these different dimensions necessarily coexist in any given moment.[20]

Since the Soviet Union launched the first Sputnik on October 4, 1957, satellites have played a central role in elucidating the shape of earth. Their deployment contributed in particular to a redefinition of the relation between the geoid and the ellipsoid. Jerome D. Rosenberg, the director

FIGURE 5.2. *Earthrise*, a photograph taken from space on December 24, 1968, by Apollo 8 astronaut William Anders, gained iconic status in the late 1960s. It came quickly to symbolize the planetary dimension of many human struggles, and the fact that environmental issues know no boundaries. The apparent fragility of the planet amid the darkness of the universe was taken as a call for humans to cooperate.

of NASA's geodetic satellites program, allegedly stated in the mid-1960s that, since the launch of Sputnik, "our knowledge of the form and gravitational field of the earth has increased more . . . than during the previous 200 years." Almost immediately data gathered by specialized satellites were used as proxies for geodetic measurements, primarily through the observation of "perturbations in [their] movement." Detailed lists of the often contradictory requisites for explicitly geodetic satellites were soon published, and as early as 1961, a few experimental satellites dedicated to geodetic tasks were launched. In 1958, the same year the United States put into orbit its own first satellite, multiple articles were published analyzing the Sputnik launch and discussing the possibility of using satellites to refine observations of the planet's gravitational field, and thus of the geoid. In 1966, the geophysicist William Kaula published his seminal *Theory of Satellite Geodesy*, which included a detailed preliminary analysis of artificial satellites' orbit perturbations.[21]

In search of methods to broaden a burgeoning global database of gravitational measurements, geophysicists came across another community of scientists, specialists in satellite remote sensing, who were eager to find practical issues to which to apply the techniques they were developing. From the beginning, satellites showed potential to overcome the undersampling and underestimation of temporal changes that had characterized earlier data gathering, as well as the vagaries of geopolitical control caused by hot and cold wars. Giff Ewing, a US pioneer in the field, noted how, following the International Geophysical Year, oceanographers had increasingly come to appreciate the difficulties inherent in averaging height and gravity measurements between stations set up mainly for the needs of a land survey and widely dispersed in both space and time. The chance to collect data from beyond the limited number of existing stations, reaching from orbit to the high seas, promised to revolutionize the way in which sea levels were conceived.[22]

In the late 1960s and early 1970s, both the United States and the Soviet Union actively pursued ways to improve marine geodesy by means of satellite technology. This was part of a generalized push to improve geodetic methods and techniques. Besides planning a system of high-altitude satellites dedicated to determining coordinates—which some ten years later would become GLONASS, the Soviet satellite navigation system—Soviet geodesists envisioned using low-altitude satellites to investigate the earth's gravitational field on the basis of variations in the satellites' movements. Combining this information with data from satellite altimetry would allow researchers, it was hoped, to improve the precision of their measurements of the satellite's height above the planet's surface from a few meters to just one.[23]

The 1980s were marked, at least in the United States, by a push toward greater understanding of the sea, justified by its possible economic and military uses. And satellites offered the possibility of overcoming the limitations of collecting data at sea by boat. Their use seemed particularly apt for studying the shape of the geoid, as it allowed researchers to assess the intensity of gravity directly on the surface of the sea. In the early 1980s the accuracy of gravimetric measurements was further improved by combining them more strictly with altimetric observations. The products were, however, general mappings of the geoid—that is, rough estimates of where the surface of the sea ought to be—rather than detailed investigations of actual water levels. The resolution was not yet detailed enough to allow more specific analyses.[24]

In 1993, with the launch of the French-American TOPEX/POSEIDON satellite, the first dedicated oceanography satellite, researchers gained a tool to continuously monitor global mean sea level from orbit—and to overcome the historical bias of tide-gauge observations. This satellite has since been complemented by NASA's Gravity Recovery and Climate Experiment (GRACE) and ESA's Gravity Field and Steady-State Ocean Circulation Explorer (GOCE) projects, which combine altimetry and long-wavelength gravimetry. Obviously, as with every technological advance, these developments affect only future data. Analysts still must take into account the intrinsic biases of past measurements, as well as technological and methodological developments. The shift in the early 1990s from tide gauges to satellites as the primary way to record global mean sea level has been invoked, for instance, by climate skeptics as a reason to doubt scientists' capacity to compare rates of change across technological periods. Might the increased rate of sea-level rise recorded since the 1990s, they ask, be a product of the tools used rather than the index of an actual geophysical phenomenon? In truth, tide-gauge data have continued to be recorded in the years since the launch of TOPEX/POSEIDON. A comparison of the parallel series shows, if anything, that between 1993 and 2013 satellite altimetry has produced lower estimates of global mean sea-level change than have tide gauges.[25]

Satellite oceanography, in both its altimetric and gravimetric components, appears to offer the absolute precision, objectivity, and scientific neutrality in determining sea levels that eluded earlier scholars and practitioners. In about twenty years, from the launch of GEOS-3 in 1975 to that of TOPEX/POSEIDON in 1993, the accuracy of measurements was said to have improved tenfold. Subsequent satellites made it possible to establish the height of the sea with yet greater accuracy. Satellites have allowed scientists to overcome the limits of a sparse network of tide gauges and gravimeters, along the coasts and on midocean islands, finally supplying continuous—and nearly global—sea-level data. The estimated average margin of error of the most recently deployed oceanographic satellite, OSTM/Jason-3, is (in continuity with its predecessor, Jason-2), 3.3 centimeters, and the aim is to improve it to just 2.5 cm, double the estimated accuracy of TOPEX/POSEIDON. In the current century a new generation of satellites combining geodetic and oceanographic observations seems to have opened new pathways to redefining height reference systems.[26]

Yet satellites' precision is not absolute. And the usefulness of the coastal measurements so criticized by German geodesists in the nineteenth century appears vindicated. Only by comparing satellite data with historical series produced at tide gauges were researchers able to fix a calibration error that was causing satellite instruments to show a slower than forecast trend in sea-level rise. In fact, valid, long-term series of observations reliably collected on the ground are essential for calibrating the instruments deployed on satellites. As noted by historians of science and technology Sabine Höhler and Nina Wormbs, the new, more-sensitive instruments are also more complex and prone to error.

Like the observers dispatched to laboriously record sea-level fluctuations on the simplest tide gauges along the coasts of Europe in the early nineteenth century, the hardware and software used to collect and assess measurements in the twenty-first can be subject to drifts and errors, which can carry significant consequences. Satellites allow us to get a sense of global sea level thanks to the mass of observations they collect and their ability to cover the whole globe without apparently favoring one area over another. But in reality they remain dependent on the availability of local, land-based observations and long-term data series. And their history is interwoven with that of Cold War surveillance, which may undercut their purported neutrality and objectivity. The choice of instruments deployed on satellites determines, as noted again by Höhler and Wormbs, what is recorded: strategical and tactical decisions have played crucial roles in how satellite altimetry and gravimetry developed since the launch of the first Sputnik.[27]

METRIC OF CHANGE

In parallel to the steady improvements of satellite technologies, earth scientists in the late twentieth century developed new ways of looking at change. And in this context sea level became an ever more relevant baseline for evaluating global anthropogenic changes. By 1975 the possibility that the recorded trend of atmospheric cooling might be reversed due to increased emissions of carbon dioxide had begun to register with climatologists. The geochemist Wallace Broecker, in what is said to have been the first article to use the phrase "global warming," noted the extreme likelihood that anthropogenic climate change would affect the level of the sea in unexpected ways. This idea had immediate success and bolstered research on the connections between increases in global sea level and in mean global temperature.[28]

As early as 1982, Robert Etkins and Edward S. Epstein of the US National Oceanic and Atmospheric Administration (NOAA) highlighted the value of global mean sea-level variations as significant indicators of climate change. The primary factors in climate-related sea-level rise, they noted, were thermal expansion and deglaciation. Etkins and Epstein also suggested that melting is an endothermic process—when solid ice changes state to liquid water, heat is absorbed from the atmosphere. The water contributes to sea-level rise, while the cooled air offsets some of the global increase in temperature. They called for a combined effort that would monitor not only sea levels, ocean surface temperatures, and the planet's speed of rotation, but also, thanks to progress in satellite altimetry, the polar ice sheets.[29]

The following year, a report from the US Environmental Protection Agency (EPA) made headlines, with dire forecasts for the effects of global warming-induced sea-level rise on coastal cities and island nations. "As early as 2100," the *New York Times* science editor wrote, "New York City's climate could be like that of Daytona, Fla. Preventive action, the report said, should be taken now." Leaning on a less pessimistic report by the National Academy of Sciences, US president Ronald Reagan's science advisor dismissed the EPA report as "unwarranted and unnecessarily alarmist." Despite the activist anti-environmentalism of the Reagan administration, the growth of the debate and the involvement of government agencies contributed to a sensible increase in the rhythm with which studies of global sea-level change were undertaken.[30]

Public awareness of sea-level rise grew, as did its urgency as a political issue. In 1987, the US Congress debated how humans contribute to it, while the Canadian minister of the environment conceded the possibility of coastal floods due to the greenhouse effect.[31] In November 1989, the Dutch Ministry of Environment organized and promoted the first intergovernmental conference on climate change. The Dutch had a historical and pragmatic interest in anything that might affect sea level, since a good part of the Netherlands has been reclaimed from the sea and lies at, or even below, its level. Environmental activists intervened at this event to press for an agreement to radically curb carbon emissions. Among other actions, they organized a press conference with the minister of the environment of the island nation of Kiribati:

> There is no place on Kiribati taller than my head, began the minister, who seemed barely more than five feet tall.
>
> So when we talk about one-foot sea-level rise, that means the water is up to my shin.
>
> He pointed to his shin.
>
> Two feet, he said, that's my thigh.
>
> He pointed to his thigh.
>
> Three feet, that's my waist.
>
> He pointed to his waist.
>
> Am I making myself clear?[32]

Bolstered by growing public interest, research efforts to assess trends in sea-level rise flourished in the late 1980s. In 1988, for example, the Japanese National Institute for Environmental Studies looked at connections between the greenhouse effect and sea levels. A year later the Australian government promoted an international initiative to establish a Pacific network of stations dedicated explicitly to measuring sea-level rise.

Another issue that garnered interest beyond the scientific community was the impact of rising oceans on maritime boundaries. In the extreme case, entire countries like Kiribati might disappear. But other ramifications also stirred debate within the legal community. According to the United Nations Convention on the Law of the Sea (UNCLOS), the normal baseline for a nation's maritime claims is the low-water line along its coasts. Any change in sea level could thus change the extent of territorial waters. In practice, such changes are neither automatic nor instantaneous; the convention leaves the choice of the charts to be used in determining sea boundaries to each individual country, so change depends both on the speed with which a country updates its charts and on political decisions regarding whether to recognize the new charts as official.[33]

In 1990, the first scientific report of the Intergovernmental Panel on Climate Change (IPCC) included a chapter on the connection between sea-level rise and climate change, further proof that the issue had, since the 1970s, gained extensive recognition. The synthesis presented by the report's authors, in line with much earlier works by Thorarinsson and Gutenberg, still assessed the past impact of thermal expansion and deglaciation on sea levels at about 10 centimeters over a century. But looking forward, the report foresaw, under a "business-as-usual" scenario, a rise of between 8 and 29 centimeters in the following forty years. Over eighty years the authors predicted a rise of between 21 and 71 centimeters—a marked acceleration in the rate of change. The concept of global mean sea level—that is, an average of averages—was crucial to the effort to use sea-level rise as a metric of climate change. All the studies cited in the IPCC's first assessment used the same, geographically biased dataset: the tide-gauge observations collated by the Permanent Service for Mean Sea Level over almost sixty years. While PSMSL represented an astounding effort of global cooperation, its planetary reach was limited by the fact that its data favored the Global North, a result of the impact of colonialism on the geographical distribution of tide gauges. As the IPCC acknowledged, "The geographical bias inherent in any global dataset will propagate into all studies."[34]

Over the course of the twentieth century, international projects to collect and compare sea-level and climate data, along with advances in satellite technology and computation, revolutionized both marine and terrestrial geodesy. This had a major impact on how the reference frame for elevations is defined. In particular, the development of space geodesy allowed practitioners to rethink how to assess heights on land and how measurements made at distant points—in both space and time—are related. In a sense, satellites have made real the idea that mean sea level can be global. Enough data can now be assembled to compute averages on a planetary scale.

Satellites have also made it possible, after centuries of doubt about its viability as a standard, to abandon sea level as a reference plane. Since the late 1970s national survey agencies and international science consortia have developed new reference planes of very different kinds. Some have focused on the geoid, emphasizing direct measurements of the gravitational field. Others went back to broad mathematical generalizations, such as the ellipsoid adopted by the World Geodetic System in 1984. The North American Vertical Datum of 1988 (NAVD88) ended up being based upon only one set of tidal gaugings—in Rimouski, Quebec, Canada. Further work, being done jointly by the United States and Canada, will average data from two hundred coastal stations to determine a new continental vertical datum and improve the geopotential model and its approximation to the ellipsoid. The United European Levelling Net (UELN), first established by the IUGG in 1955—a belated conclusion to the late nineteenth-century debates about the unification of the European vertical reference system—no longer provides metric elevations referred to sea level, but rather geopotential heights calculated on the basis of gravitational attraction. In a somewhat ironic turn, its primary reference point—proposed already 150 years ago—is the mean high tide level recorded over almost five centuries in Amsterdam.[35]

The relatively short data series produced to date by satellites cannot tell us much about global sea-level change if they are not considered in relation to variously discordant, local observations. Sea level exists concurrently on a multiplicity of scales, from the local to the planetary, none of which can be understood and represented in full without considering them all. Sea level is not only a local measurement of tidal movements, a blip in the orbit of a satellite, or a global aggregate of data. It is all of these at the same time.

SIX

THE RISING TIDE

Modern cartography has shaped a collective idea of the world's coastlines: most of us have clear, if rough, expectations of the boundaries between land and sea. But while local changes to the shape of coasts are nothing new, anthropogenic sea-level rise is expected to transform shorelines to an extent that is incomparable with earlier events. Changes now happen faster and on a global scale. Just as mapping produced our shared image of the world, it now plays an essential role in visualizing how this chronological acceleration and geographical expansion will affect land-formation processes. In the last decades—possibly beginning with the famous graphics in Al Gore's 2006 documentary *An Inconvenient Truth*, which showed, for instance, a submerged Florida and a flooded Manhattan—maps showing the impact of sea-level rise on shorelines have been ubiquitous.[1]

A simple search on Google brings up multiple interactive maps that allow one to visualize how different increases in the level of water might reshape the world's coastlines. The principle is simple: the user selects an increase in sea level (or some other relevant variable), and a color layer, usually blue, is overlaid on land below certain elevations. The visual impact is striking and worrying on the global scale, but of limited use if we zoom in to places we know. Like all two-dimensional representations of three-dimensional phenomena, these maps are imperfect tools. In some cases, the overlays are based solely on elevation, whether or not the land they cover is next to a coast where seawater really could overflow it.

The oldest, still-working interactive map I have found was set up in 2006, the year Gore's documentary was released, and is based on NASA data. It offers a simple and effective visualization of how radically climate change might transform the coasts we have come to know through the apparent stability of the Holocene, but it does so only for extreme changes: while it will simulate a sea-level rise of up to 60 meters—the forecasted impact of the melting of all of Antarctica's ice cover—there is no option to see the effects of plausible, short-term rises of less than a meter. Nor does it offer ways to translate modeled timelines or increases in temperature into sea-level change. Its author acknowledges these and other issues with the tool. For one, it adopts mean sea level as its reference plane and does not consider local tidal ranges, the effects of coastal erosion, or the role of existing coastal defenses, all crucial factors when considering the impact of sea-level change on the aspect of coasts.[2]

While global mean sea level provides an easy way to smooth out local particularities, rising water levels and the decoupling of geodesy from mean sea level have brought high tide lines back to prominence in dis-

FIGURE 6.1. Tools intended to map potential changes to coastal landscapes due to higher sea levels are widespread on the internet, at the latest since such maps appeared in Al Gore's 2006 documentary *An Inconvenient Truth*. This screenshot represents the projected impact of a one-meter rise in sea level on the coasts of northwestern Germany, the Netherlands, Belgium, northeastern France, and England, according to Alex Tingle's Flood Maps website.

cussions of the boundary between land and sea. In recent years, newer, more subtle mapping efforts have become available. One set up by the climate news agency Climate Central, for instance, discards mean sea level in favor of high water—mostly estimated rather than measured. Its tool thus offers a more nuanced approximation of the actual impacts of sea-level rise on the world's coastal landscapes. Furthermore, it allows for much smaller increments of sea-level change, making it theoretically possible to visualize how littoral spaces will look in as little as ten or twenty years. And it offers multiple visualization criteria: users are able not only to view the impact of determinate linear increases in the level of water, but to specify an increase in global temperature or degree of loss in the ice sheets and calculate how the consequent sea-level rise would affect coasts. Possibly the most interesting feature is one that allows users to compare the differential impact of future climate scenarios—which reveals that the most extensive and transformative impact on the coasts appears already in the scenario corresponding to global warming of 1.5° Celsius above preindustrial levels.

Like other global depictions of change, the tool cannot, however, realistically track potential fluctuations at the local level. Its modeling fails, for instance, to account for already existent flood defenses: even with sea-level rise set to zero, I found, the tool presents most of the Netherlands— substantial parts of which are indeed lower than the sea level—as submerged. It appears, though, to effectively avoid highlighting inland areas as at immediate risk. Neither the area around the Dead Sea, the north coast of the Caspian Sea, nor the Tunisian salt lakes are marked as affected by any sea-level rise of less than 10 meters; the older tools all mapped these regions as flooded by a rise of even a single meter. Other tools, like the NOAA Sea Level Rise Viewer, unfortunately limited to the contiguous United States, have overcome this issue by mapping only coastal areas.[3]

APPREHENDING SEA-LEVEL RISE

According to historical tide-gauge data, the twentieth century experienced an average yearly increase in global mean sea level of 1.7 millimeters. Recent satellite observations show that this rate has more than doubled in the twenty-first century, to 3.6 millimeters. Overall between 1900 and 2017, the sea rose by up to 21 centimeters, about a third of that after 1993. Models predict an even higher rate of increase for the next decades, projecting, relative to the level recorded in 2000, a rise of at least 15 centimeters by 2050 and possibly more than a meter by 2100. Some recent studies foresee, under high-emission scenarios, a rise of up to 2.4 meters. And all models essentially concur that sea level is extremely likely to continue to rise beyond the end of this century. The prospect rightfully stirs dread. The world of the future will look radically different from the one humanity has known through most of the Holocene. But, as exemplified by the simple tools discussed above, models and forecasts don't yet provide much specificity about what it will mean for affected coastal territories.[4]

The question of how to model and visualize future variations in sea level and their impact on global coastlines reminds us, once more, that definitions of sea level are technological and social constructs, as are representations of its changes. Different sea levels, and different ways of assessing change, are adopted for different uses, on scales from the global to the extremely local, involving measurements deemed absolute or relative. Like many other elements of human understanding of the environment, sea level has been shaped as part of a process termed "environing." How sea level is understood and perceived depends on the technologies and media used to register and describe it. Moreover, the materiality of sea level is much blurrier than boundaries on maps and numbers in tables would allow us to think. Scientific processes of simplification allow for the production of understandable representations of extremely complex systems; global mean sea level and relative sea level are just two of many examples. According to the political scientist James Scott, these processes are part of a more general effort by modern societies to make the world legible. As we face the evident impacts of climate change on the global ocean and its shorelines, such simplifications will play an increasingly important role, for example, in the production of borders. Scientific averages, produced for certain limited uses, are translated ever more frequently into the basis of legal judgments.[5]

International courts seem to agree that changes in the actual coastlines of a state will impact the extent of its territorial waters and exclusive economic zones—the marine areas in which it has special rights of exploration and resource use. Were sea level to rise by 1.5 meters, for instance, the Florida coastline would recede and various archipelagic baseline points in the Bahamas could be submerged. Cuba's coast is, on average, steeper and would be relatively less affected by the rising water. As a result, Cuba could assert that its maritime zone—based, under agreements with the United States and the Bahamas, on a principle of equidistance from each country's baseline—should be enlarged. For their part, the US and the Bahamas would likely argue for overcoming the equidistance principle. Similar conflicts might well be ignited in many places around the world. In the Philippine Sea the same 1.5-meter rise would completely submerge the Okitonorishima atoll, which Japan asserts is an island, claiming an exclusive economic zone around it. Chinese authorities hold, to the contrary, that it is just a rock, with no potential to sustain human life. The atoll's submersion would put an end to Japanese claims to the surrounding area's rich natural resources, which could then be exploited, and eventually exhausted, by competing nation-states.[6]

Throughout the eighteenth century, the level of the sea was mostly seen as a range encompassing incomparable local tidal extremes, and its fall was considered an explanatory factor for differences measured over time. In the early nineteenth century, the sea was reframed as stable and its potential as a reference point institutionalized. In this process, its level was redefined as an average: mean tide level. The multiplicity of local "mean sea levels" made possible by the development of automated tide gauges and increased investment in survey work were generally assumed to coincide—to be effectively at the same altitude at any given point along the world's coasts. In recent times, geodesy has developed new methods, independent of the gauging of the sea, to ascertain heights.

The "mean rate of global averaged sea level rise" is the formal terminology adopted by the IPCC to describe what is known in common parlance as sea-level rise. It is a construct—both *global* and *averaged*—made possible by the collection of a huge number of observations. In the early nineteenth century, the averaging of local measurements was made convenient by the development of automated tide gauges. In more recent times, the sheer quantity of data available to ascertain global averages has boomed due to a rapid increase in observations collected locally along the world's coasts, the first gravity measurements at sea, the deployment of satellites, and the exponential growth of computational power. The longest series of such data were produced in the context of measurements undertaken to determine national vertical datums.[7]

THE MEANING OF DATUMS

As mentioned at the end of chapter 5, mean sea-level datums are continually being recalculated. Moreover, alternative ways to assess elevation, referring not to the sea but to geometric approximations of our planet's shape, have found new uses since the late twentieth century. Ever more accurate ways to ascertain heights—in recent years, through the assessment of gravitational fields—are being developed. A study from 2010 proposed comparing the speeds at which optical atomic clocks tick at different altitudes, a technique, applicable over great distances without major losses in precision, that shows great promise for establishing geodetic datums freed from both a material connection to the sea and the limits of a geometrical approximation.[8]

All of this sounds impressive and, in principle, very much desirable. But people have been trained over the last couple centuries to think of heights in relation to the sea. Especially if they live along a coast, they are interested in where their homes are placed in respect to water. National and international reference planes may be confusing since they do not reflect local perceived elevations, and this is even more true for these new geodetic datums. For instance, the ellipsoid, the approximate geometrical surface of the planet, is an excellent reference plane for satellites. But for practical, local uses, "geodetic heights are not physically very meaningful," notes the Canadian geodesist Petr Vaníçek. They may, in fact, differ from local measurements by up to 200 meters. "Height consumers," Vaníçek says, have become accustomed to thinking of sea level as the "natural" zero.[9]

In April 2016, en route from one conference to another to present preliminary results of the work I was doing for this book, I stopped by the NOAA library in Silver Spring, Maryland, to do some archival research on the measurement of mean sea level in the US. While there, I spoke with Daniel Roman, chief geodesist of the US National Geodetic Survey, about the choice of national and global reference planes and attempts that had been undertaken to minimize the impact of the difference between sea level and the ellipsoid on life in coastal areas. Sitting in a corner of the library, by a window overlooking Silver Spring's dense downtown and metro stop, we discussed in particular how global projects to define an international reference system still relate to national efforts. The geodetical surveys of the USA and Canada, for instance, are working to define a new common reference plane that will take into consideration both the earth's gravitational field and sea-level gaugings, and thus combine the global and local dimensions. The quest for international reference frameworks isn't yet finished and, as was the case more than a century ago, still depends on the goodwill and participation of national agencies. Roman stressed, in fact, how new global reference planes always need to be defined in conversation with both local measurements and historical series. For example, for a homeowner living along one of the long, sandy beaches in the Carolinas, which, like much of the US Atlantic seaboard, have low-lying coastal zones stretching inland for miles, it would not make much sense, he noted, if the height benchmark lay a meter or more above the perceived local level of the sea—which might well happen were one to impose global averages at the national and local levels.[10]

At each scale, different benchmarks are used. These points are put into conversation when comparisons are deemed necessary, an approach uncannily reminiscent of that proposed by Lallemand in 1889 and discussed in chapter 4. As the historian Dipesh Chakrabarty has recently noted regarding climate science, translations between the local and the planetary are a necessity of most processes of environmental knowledge production. The materiality of the local must constantly be put into dialogue with the abstract dimension of a planetary science.[11]

The low expanses of the Atlantic seaboard are home to more than sixty million people, many in densely populated conurbations, and to some of the US defense forces' biggest military installations. There are thus heavy investments in housing, industry, and transportation. The region is, as well, notably vulnerable to hurricanes, flooding, and storm surges, threats that make it and the Gulf Coast into hotspots when it comes to possible impacts of ongoing sea-level rise. Anthropogenic sea-level rise affects littoral regions directly—through erosion, the flooding of some settlements, and day-to-day changes in the aspect of the coastline—but also indirectly, by contributing to the intensity of hurricane storm surges: the higher the mean water level, the higher and more far-reaching the possible floods. And it affects different social groups living along the coasts differently, exacerbating inequalities. The poor are worse off than the rich, and it seems that race trumps even class in determining the severity of the consequences of an extreme weather event. Hurricanes and superstorms in the region have increasingly hit the most vulnerable communities.[12]

WAITING FOR DISASTER

Hurricanes and other storms are among the more immediate of the many possible effects of long-term processes unleashed by climate change. But even at an increased rate of sea-level change, like what we've experienced in recent decades, conceptualizing the sea as a mean smooths out the perception of danger. The long timeline of change also helps hide some of the risks to which coastal populations are exposed. Sea-level rise is happening; there is little doubt about that. While not always noticeable, a tipping point has already been reached, and the augmented impact of climate-induced events is increasingly evident. What makes sea-level rise problematic as a baseline in this context is its gradual nature; slow but continual changes in water levels can be hard to identify, which has contributed to the notion of the seas' long-term stability. The driving force of a slow disaster, sea-level change hides its severity behind time.

Disasters, as the historian of technology Scott Gabriel Knowles notes, are always deeply historical constructs, processes that depend not only on incidental or natural causes but on the political, technological, and social settings within which they are embedded.[13] In an essay on luxury real estate in a drowning Miami, journalist Sarah Miller quotes a real estate agent:

> "I mean, it's not like, you're going to wake up one day and the ocean is outside your window!" We laughed again. "*I think* you know, maybe, something—if something is gonna happen it's like, I think, it's like, 100, 200 years."[14]

But something is happening, even as the realtor speaks. The cost of flood insurance is predicted to increase fourfold due to sea-level rise and its side effects. Miami's groundwater is rising as well, contributing to a quadrupling over the last fifteen years of what has been termed "sunny day flooding"—sudden increases in water levels in urban settings without immediate meteorological causes, due more to tidal extremes than the average signaled by mean sea level. Technological fixes like pumping stations and raised streets give an impression of safety, as Miller observed in some of her interactions with real estate agents. But as Miami, dubbed the "most vulnerable" major coastal city in the world, steadily sinks, there is no comprehensive plan. The interventions are small, piecemeal, not coordinated with broader, citywide processes or even with the steps taken by one's neighbors. The rich move inland and upward, acting as individuals, unconcerned with the preservation of a functional urban infrastructure, while around them other communities, more dependent on a working social system, face pressure from developers seeking to remove them from the high ground to which they were forced during the city's earlier waterfront boom, or find their own homes and savings threatened due to sea-level rise.[15]

This waterfront class struggle is a global phenomenon. During a visit to India's Nicobar Islands—a low-lying archipelago in the eastern Indian Ocean, at risk of becoming uninhabitable due to sea-level rise—in the aftermath of the tsunami that struck in 2004, the Indian writer Amitav Ghosh noticed how the richest, in a reversal of usual trends, had been the hardest hit. As in Miami, waterfront properties had been the most coveted, suggesting a belief "that highly improbable events belong not in the real world but in fantasy." Modern ideas of holocenic stability, expressed through a colonial vision that attributes positive qualities to the sea, have become deeply embedded in the contemporary worldview. The desire to live on the coast, looking out on the infinite expanse of the oceans—as opposed to distancing oneself from the water's dangers—is a relatively recent cultural development, approximately coeval with the conceptualization of sea level as an average and of the sea as stable.[16]

As Ghosh notes, "'Growth' in many coastal cities around the world now depends on ensuring that a blind eye is turned toward risk." The shoreline has become prime real estate, the location of the most expensive new developments. But as we have seen in the case of the US Atlantic coast, even the slowest disasters can rapidly shift gears and break out in bursts of violence. The risk of cyclones is increasing due to climate change along the western coast of India, a region that has historically experienced few such extreme weather phenomena. Once the sea rises by a meter, large parts of South Mumbai, a low-lying tongue of land, mostly reclaimed from the sea, will become again uninhabitable.[17]

RECOVERING OLD METRICS

If colonial conquest brought with it a reconsideration of humanity's relationship with the sea, it also imposed a stricter separation between land and water, as discussed in the introduction. But porous spaces, in which land and sea mingle, are resilient and, beyond the nominal distinctions made by states and international agencies, still common throughout the world. Bangladesh, one of the poorest countries of the planet, is particularly prone to high-amplitude floods, to the point that only a flood that covers more than half of the land is considered "heavy." In the Netherlands, among the richest countries, proposals have been made to embrace higher sea levels and revert to premodern models of cohabitation with the sea, favoring managed flooding over dikes and concentrating human communities on isolated mounds surrounded by water.[18]

In 1970, Bhola, the deadliest tropical cyclone on record and one trigger of the secession of East Pakistan and the birth of Bangladesh, caused a huge mass of river silt to precipitate into an island in the Bay of Bengal, in front of the Sundarbans, the world's greatest mangrove swamp. This accretion of fluvial discharge, in seasonal flux, oscillating between 2 and 12 square kilometers depending on tides and rainfall, soon became an object of contention between India and the newly created country. Never settled but formally declared an island according to the law of the sea, it was claimed by both countries as part of their territorial space—called South Talpatti in Bangladesh and New Moore in India. Except for some tiny bits that resurface at extremely low-tide conditions, well below mean sea level, the island disappeared as the water level in the bay rose in 2010, before an international court could arbitrate the case. Yet this was only one of about a dozen "vanishing islands" off the Ganges-Brahmaputra delta system, characterized by their extreme sensitivity to even minimal sea-level changes. Another such island, Lohachara in India, disappeared beneath the encroaching sea in 2006. Lohachara, however, was inhabited, and the six thousand residents became environmental refugees, making the story a cautionary tale for what could happen to vast swaths of the coast of the Bay of Bengal.[19]

The threat is not limited to these flimsy islands. The whole of Bangladesh is vulnerable, as well as much of India's coast. A low-lying deltaic country, almost a third of whose population lives by the coast, Bangladesh is virtually defenseless against tidal floods and sea-level rise. This is not a forecast but, as the vanishing islands exemplify, a process already affecting land erosion, soil salinization, and loss of biodiversity. Geological factors make the region's situation even more dangerous: tectonic movements are producing, in parallel to the rise of the sea, a lowering of the land. The combined effect of these phenomena means that relative sea level is rising two to four times as fast in Bangladesh as in the Netherlands. Once the relative sea-level rise reaches 45 centimeters, more than 15,000 square kilometers of land will be permanently lost. Double that, and more than a fifth of the country will end up underwater, making up to fifteen million people homeless or forced to migrate. Determining mean sea level and its rise in the region is not easy, however. Research done in 1994 on the quality of Bangladesh's tide gauges led to the realization that the existing ones were of limited use; most were located within the delta and had rudimentary benchmarks, whose stability was virtually impossible to check. Moreover, repeated cyclones damaged tide gauges to the point that, from 1977 to 1987, the level of not one single storm surge was registered. And little seems to have changed since then.[20]

The Sundarbans of southwest Bangladesh have been intensely developed since the 1960s. Thousands of square kilometers have been reclaimed, creating new agricultural land just a few feet above sea level. This land is now at risk. Due to multiple local causes, including a proliferation of embankments, tidal ranges—the difference between mean high and low water—have increased in the last decades. Analysis of the ongoing processes has led researchers to suggest using *effective sea-level rise*, that is, the high-water maxima, instead of relative mean sea level as an indicator of risk. In the face of local human interventions and a growing tidal range, this benchmark could provide a more precise signal for possible dangers to land, biodiversity, and communities. As a consequence, it could offer a clearer image of the risks to which deltaic lowlands are exposed in a time of rising sea levels. It is, moreover, worth remembering that high water has historically played a crucial role in measurements: as discussed in chapter 3, the Amsterdam datum, already used as the basis for the German *Normalnull* and proposed as the common European reference plane, is determined by recording mean high water.[21]

The suggestion that effective increase in sea level may be a more precise marker is not unique to Bangladesh's Sundarbans. Alternatives to mean sea level as a baseline have come back to the forefront all over the world since climate-induced sea-level rise was first hypothesized. In a brief statement written in 1981 for the Rosenstiel School of Marine and Atmospheric Sciences at the University of Miami, the geoscientists Harold Wanless and Peter Harlem adopted an observation reminiscent of the upper limit of algal growth used in Venice as a benchmark for depth soundings. The idea was to complement existing data series with alternative, more material and haptic metrics. Wanless and Harlem compared recent images of barnacle and oyster incrustations on sea walls with pictures collected by the marine ecologist Hilary Moore more than three decades earlier. In line with earlier studies proposing that barnacle colonies could be used to assess historical changes in the absence of instrumental gaugings, they ascertained that in the waterway passing Coral Gables, south of Miami, the upper limit of the two species' vertical range was up to 16 centimeters higher in 1981 than in 1949. While approximate, the changes assessed through this animal marker—which falls, on average,

FIGURE 6.2. Just as algal growth has been used historically as a marker of sea level in Venice, scientists have used the vertical level reached by barnacles and oysters growing on a bridge in Coral Gables, Florida, as seen in photographs from 1949 and 1981, as a proxy for changes in sea level. *Source:* Wanless and Harlem, "A Statement on the Evidence for and Implications of a Recent Rise in Sea Level." Courtesy of the authors.

a bit short of annual mean high water—correspond to the sea-level rise measured since the Great Acceleration of the 1950s. The method, while partly tied to local conditions, had an additional advantage: the permanence of the animal marker means the paired photos can be meaningfully compared even if they might not be taken at the same time of the tide.[22]

EMBRACING THE RANGE

To underscore the risks posed by climate change-induced sea-level rise to small island states and their populations, Simon Kofe, Tuvalu's foreign minister, recorded a video to be broadcast at COP26, the 2021 UN Climate Change Conference in Glasgow. Standing at a lectern in the middle of the sea, wearing suit coat, tie, and shorts, he calls on world leaders to act quickly to help the countries most vulnerable to the effects of rising sea levels. Twelve years earlier, in advance of COP15 in Copenhagen, Mohamed Nasheed, then president of the Maldives, staged an undersea cabinet meeting—he and his ministers, clad in wetsuits, discussed matters of state some four meters underwater in a lagoon of their archipelago—to send a similar message to the world. Time goes by, but the desired impact of such messages does not seem to materialize.[23]

The rise of the oceans is a global phenomenon; its pace and intensity, however, vary in different parts of the world. The slow, erosive impact of tides and sea-level changes on England's coasts was, as discussed in chapter 4, already the object of inquiries by a royal commission at the turn of the twentieth century. One prime example is the flooding of Doggerland—the extension of fertile land that connected Britain to continental Europe between the end of the Pleistocene and the start of the Holocene. Others are the sea floods recorded by Australian Aboriginals in some of the oldest accounts of the impact of postglacial sea-level change.

Along the coasts of England in the past century, footpaths, churches, antiaircraft batteries, and whole towns have been swallowed by the sea. This erosion is a gradual process, creeping into our lives at a slow pace, to some extent allowing for adaptive responses but hiding the true extent of its dangerousness. It is also a process that will, in the coming century, have a major impact on our image of coastal landscapes. As shown at the beginning of this chapter, mapping tools do not fully convey the physical effects of rising waters on coasts. But beyond that, there are psychological and emotional consequences. The modifications that sea-level rise will inflict on cherished coastal landscapes will undoubtedly trigger solastalgia—the distress, according to the Australian philosopher Glenn Albrecht, caused by the impact of environmental change on landscapes to which we have an affective link. The very idea of a globally menacing sea level, eradicating the naturalized image of a stable sea construed over the last two centuries, is due to cause similar anxieties.[24]

For all its limitations, including its embeddedness in colonial epistemologies, mean sea level is still a useful metric that has served humanity well, acting over time as both a baseline for elevation measurements and a powerful symbol of anthropogenic global change. But it must be kept in mind, as discussed in the previous chapters, that it is not the only possible way to visualize and understand sea level. In what might seem a drastically disjunctive interpretation, the Māori people of Aotearoa/New Zealand, for instance, don't see land and sea as in binary opposition, but rather as parts of a continuity that can only be understood holistically. Accordingly, there is no need for sea levels as they were presented by colonial geodesists. Sea level as a concept is a product of Western ontologies. But even within Western science, a true comprehension of how sea and land interact increasingly requires a multiplicity of reference points on a variety of spatial and temporal scales. Oceanographers, for instance, model the geoid as stable and refer to the variability of the seas' surface height—that is, the differences between the geoid and the recorded state of water—in respect to fixed points. While in the short term assuming such a stability offers clear advantages, ongoing climate change through sea-level rise—as highlighted by Fairbridge in 1960 and discussed in chapter 5—will also gradually affect the geoid.[25]

The practice of gauging sea levels and the choices we make as a society and as a scientific community about baselines, scales, and metrics—whose historical background was examined in chapters 2 and 3—still inform the way humanity interprets global changes in the level of the sea. In this book I have highlighted how the conceptualization of sea level as an average, an approximation, and a neat separator between land and sea, as well as the materiality of its measurement, affects our current understanding of its rise as a gradual process. I have also stressed how, as with all scientific categories, the choice of which sea level to adopt and use is contingent. A time of change asks for more nuanced interpretations, ones that accept and foster different categories of analysis at different scales. If the public accepts such an understanding of sea level, it might become possible to produce more effective political responses to the threats posed by a global rise in water levels—not because one or the other level is intrinsically better, but because each analytic framework is specifically attuned to a particular historical moment or spatially determined condition. The local and the global dimensions of future changes along the oceans' coasts must always be considered holistically, providing clear and understandable information of how the different scales and systems of reference are connected to each other. Profiling the conceptual historicity of sea level as a baseline and examining its constructedness can contribute to achieving this. At the same time, in an era in which the sea has lost its stability, the ways changes to our coasts are visualized need to be rethought, possibly retrieving interpretational models dismissed in the past, so that the actual risks are highlighted. The materiality of sea-level rise and its measurement is a crucial part of the message.

ACKNOWLEDGMENTS

As lonely as it sometimes feels, historical research is never done alone. The work behind this book is no exception. Writing it was made possible only by the support of many institutions and individuals. When the project was little more than a draft of an idea, I was lucky enough to be funded by the Deutsches Museum in Munich to explore its feasibility. For this and the productive time I spent there, I am grateful to Helmuth Trischler and all of the museum staff. Two years spent teaching at the University of Wisconsin–Madison allowed me to test some of the ideas that have become part of the book in front of exceptional students and to gather very fruitful feedback. Thanks to the German Academic Exchange Service—DAAD, Gregg Mitman, and Marc Silberman for having made my time there possible.

The Max Planck Institute for the History of Science (MPIWG) then proved to be the ideal environment to hone that research idea. Everyone in Department 3 read early versions of various chapters. Their feedback and comments contributed greatly to reframing and reorganizing my arguments. The visiting scholars of the Art of Judgement Research Cluster helped me better articulate why measurements matter. Julia Sánchez Dorado, Shu Changxue, and the other participants in the Measuring the

Earth discussion group deserve special credit at this regard. Tom Lekan and Sebastián Ureta were quintessential in setting up the Baselining Nature workshop in difficult times. The work we did at that meeting was extremely helpful in figuring out the place of reference points in recent history. My appreciation goes also to Martin Mahony, for thinking with me about the place of verticality in the history of science, and to Giacomo Parrinello and Etienne Benson, for exploring with me the place of estimation and approximation in water sciences.

The growing trend toward the digitization of primary and secondary historical sources was crucial in making this project possible. Ester Chen and her staff at the MPIWG library digitized the minutes of all the meetings of the Internationale Erdmessung held at the GeoForschungsZentrum in Potsdam. Pascal Belouin and Brent Ho did invaluable work setting up and improving the search interface for the corpus, which contributed immeasurably to the writing of chapter 3. When digitized versions of sources were not available, I had abundant support from the library in accessing hard copies. And when even this was not enough, Dagmar Schäfer, in her capacity as the director of Department 3, was extremely generous in supporting my research needs and visits to archives. A trip to the National Oceanic and Atmospheric Administration in College Park, Maryland, was particularly fruitful. Thanks to Albert "Skip" Theberge for having facilitated it.

Audiences at multiple seminars and conference sessions listened patiently to elements of this project at various stages of completion. I am extremely grateful to all for their comments. Six months spent as guest professor in the Vielfalt der Wissensformen—Diversity of Knowledge interdisciplinary study program at Humboldt University, and the seminars I offered as part of the School of Disobedience at the Volksbühne Theater, both in Berlin, allowed me to test some aspects of this research in front of a wider audience. Thanks to Christian Kassung and Georg Diez, respectively, for having offered me these opportunities.

If research is inevitably a collective endeavor, this is even more the case for writing. This book would not look as it does had it not been for Jessica J. Lee and her terrific skills as a developmental editor. Further praise has to go to Joel Score for his thorough and careful work copyediting the manuscript and to Bonny McLaughlin for the great job she did indexing. All errors or misinterpretations that remain are, obviously, my own. It was a pleasure to work with Karen Merikangas Darling and her colleague Fabiola Enríquez Flores at the University of Chicago Press, as well as with Oceans in Depth series editors Helen Rozwadowski and Katherine Anderson. Their continued moral and material support has been essential in giving the book its final shape. My appreciation goes also to two anonymous reviewers for their insightful suggestions and their enthusiasm for this project. Special thanks, finally, to Giulia and Kuno for having made this historian's work less lonely, both when my research materialized into trips to the beach and when it translated in long hours in front of a laptop.

It may be no more than a coincidence that I finished reflecting on this book on a beach near Alghero in Sardinia—where cliffs alternate with sandy beaches in a succession of verdant bays—looking upon Capo Caccia, at whose foot the karstic Neptune's Grotto offers geologists an exceptional insight into sea-level changes since the last ice age. But it feels like having closed a circle that started in a summer many years ago on a beach on the French Riviera.

NOTES

INTRODUCTION

1. Since 1954 the height of Mount Everest has customarily been given as 8,848 meters. In 2020, China and Nepal, after having each made their own trigonometric survey, agreed that the height is 8,848.86 m, here rounded up to 8,849. The elevation of Chimborazo has long been given as 6,310 m, but GPS measurements made in 2016 provided 6,263 m as a more plausible height. According to the first GPS measurement in 2001, Mont Blanc was 4,808.40 m high. Since then its elevation has been measured every two years: in 2023, the result found was 4,805.59 m. Arnu, "8848,86"; Dongo, "Le volcan Chimborazo"; Personnaz and Herenstein, "Mont Blanc Shrinks."
2. Sea-level data since 1993 can be downloaded at https://climate.nasa.gov/vital-signs/sea-level (NASA Global Climate Change, "Sea Level").
3. See Scheuchzer, *Nova Helvetiae tabula geographica*. The Steilerhorn is drawn in the bottom right corner of the map. On Scheuchzer's approximation of the mountain's elevation, see also Heyde, *Die Höhennullpunkte*, 3; Grosjean, *Geschichte der Kartographie*, 93.
4. See Middleton, *History of the Barometer*, 3–18 (on barometer's precursors), 51–52 (on Périer's experiment), 132–33 (on use of barometer as an altimeter). For the development of the barometric formula, see Feldman, "Applied Mathematics."
5. On the premodern relationship with mountain heights in continental Europe, see, for instance, Ireton and Schaumann, "Meaning of Mountains." On Incan and

Chinese representation of heights, see Gartner, "Mapmaking in the Central Andes"; Yee, "Reinterpreting Traditional Chinese Geographical Maps," 50–53; Yee, "Taking the World's Measure," 110–12. For a graphical reconstruction of the Mawangdui topographical map, see Lucarelli, "Three Mawangdui Maps."

6. On the use of time rather than linear units to measure distance in Nepal, see Fisher, *Sherpas*, 84–85. On the assessment of heights as a physical and bodily exercise, see Fleetwood, "Bodies in High Places," 497. On the use of ascent to assess heights in Europe, see Broc, *Les montagnes au Siècle des lumières*, 128; Diderot, "Montagnes," 676.
7. On the quantitative turn in the physical sciences, see Feldman, "Applied Mathematics." For the first comparative table of altitudes, see Pasumot, "Lettre aux auteurs." For a firsthand account of Saussure's measurements, see Saussure, *Relation abrégée*, 22–24.
8. On Humboldt's ascent of Chimborazo, see Fleetwood, "Bodies in High Places," 495. For Humboldt's quotes, see Lubrich, "Fascinating Voids," 164. On the increasing importance of accuracy in early nineteenth-century altitude measurements, see Fleetwood, "No Former Travellers," 32. On the idea of global verticality, see Hardenberg and Mahony, "Up, Down, Round and Round."
9. See Cajori, "History of Determinations of the Heights of Mountains"; Bowie, "Notable Progress in Surveying Instruments."
10. On the limitations of vertical reference points adopted in premodern Europe, see Cajori, "History of Determinations of the Heights of Mountains," 483, 494–97. For an early mention of sea level as a necessary reference, see Scheuchzer, "Barometrical Method of Measuring." On the conventional nature of vertical reference points, see Heyde, *Die Höhennullpunkte*, 3. On the biblical flood and early modern sciences, see Barnett, *After the Flood*, esp. 188–95. On early modern theorizations of a globally dropping sea, see Rudwick, *Bursting the Limits of Time*, 172–80.
11. See Ureta, Lekan, and Hardenberg, "Baselining Nature."
12. Siegert, "Cacography or Communication?" 30.
13. For an overview of recent work on the place of verticality in the history of science, see Hardenberg and Mahony, "Up, Down, Round and Round." For the literature on environments made by human interventions, see Sörlin and Wormbs, "Environing Technologies"; Wickberg and Gärdebo, "Where Humans and the Planetary Conflate."
14. On the choice of Greenwich longitude as the prime meridian, see Withers, *Zero Degrees*. On the history of high-tide measurements, see Woodworth, "Tidal Measurement." On mean sea level as a temporal average, see Sammler, "Rising Politics of Sea Level," 8.
15. On the arbitrariness of reference points, see Theberge, "150 Years of Tides," 1. On the elusiveness of the sea and deceptiveness of coastlines, see Carson, *Edge of the Sea*, 1; Marmer, "Purpose of Tide Observations," 167–68. On coastlines in Western cartography, see Carter, *Dark Writing*, 9. On land-water separation as a legal construct, see Bhattacharyya, *Empire and Ecology*, 201–4 (quotation, 202). For the history of tide science, see also Cartwright, *Tides*.

16. Cazenave et al., "International Space Science Institute (ISSI) Workshop," 1.
17. On the gradualness of geological changes, see Chakrabarty, *Climate of History in a Planetary Age*, 7. On the need to historicize planetary processes, see Nail, *Theory of the Earth*, 4–11. On holocenic stability and its disruption, see Magnason, *On Time and Water*, 117–20.
18. For a sense of the available tide-gauge data, see Holgate et al., "New Data Systems"; Permanent Service for Mean Sea Level (PSMSL), "Tide Gauge Data." On gravitational attraction and the distribution of seawater, see Mitrovica, Gomez, and Clark, "Sea-Level Fingerprint"; Gudrais, "Gravity of Glacial Melt."
19. On Qiantang, see Moule, "Bore on the Ch'ien-t'ang River"; Cartwright, *Tides*, 16–18; Needham, *Science and Civilization in China*, 3:483–94. On the tsunami stones, see Buhrman, "Remembering Future Risk"; Hunchuck, "Incomplete Atlas of Stones."
20. On the history and heritage of the *comune marino*, see Rossi, "Il comune marino," 42; Zwingle, "Watermarks"; Crouzet-Pavan, "Venice and Its Surroundings," 36–38. On its limitations and the makeshift solutions devised to counter them, see Hardenberg, "Making a Stable Sea," 137.
21. On the history of the *Amsterdam peil*, see Weele, *De geschiedenis van het N.A.P.*; Dam, *Van Amsterdams Peil*.
22. For Helmert's remark on the sea's instability, see Hirsch, *Verhandlungen der . . . Permanenten Commission* (1892), 148. Moray's proposal can be found in Moray, "Considerations and Enquiries Concerning Tides," 299–300. For context and further details, see also Pouvreau, "Trois cents ans de mesures marégraphiques," 57–63.
23. On the collection of sea-level data in Brest, see Pouvreau, "Trois cents ans de mesures marégraphiques," esp. 68 for a description of the new gauge. For more on Descartes, see Aiton, "Descartes's Theory of the Tides." The Académie protocol can be found in Goüye and de la Hire, "Memoire de la maniere d'observer." On the British state's earliest efforts, see also Reidy, "Gauging Science and Technology," 5–6.
24. On "geological agency," see Chakrabarty, "Climate of History: Four Theses," 207.

CHAPTER ONE

1. For Bessel's thought experiment, see Bessel, "Ueber den Einfluss der Unregelmässigkeiten der Figur der Erde," 270. On his life and contributions to mathematical geography, see Kirrinnis, "Friedrich Wilhelm Bessel."
2. See Sieger, "Seenschwankungen und Strandverschiebungen"; Hardenberg, "Making a Stable Sea." On the eighteenth-century "standard model of falling seas" and its diffusion among European savants, see Rudwick, *Bursting the Limits of Time*, 172–80.
3. Playfair, *Illustrations of the Huttonian Theory*, 446. Playfair is known for his role in popularizing the revolutionary, if rather tedious and obscure, work on geological processes of the Scottish geologist James Hutton.
4. On the Baltic tidal range, see SMHI, "Tides"; Medvedev, Rabinovich, and Kulikov, "Tides in Three Enclosed Basins." For the quote, see Lyell, "Bakerian Lecture," 29.

5. Celsius, "Anmärkning om vatnets förminskande."
6. See Desmarest, "Ferner," 142–45; Ekman, "Concise History," 359. On Wijkström's methodology, see "Utdrag af Kongl. Vetenskaps Academiens Dag-Bok." On Frigelius's contribution, see Bruncrona and Hällström, "Beobachtungen und Angaben," 322.
7. See Playfair, *Illustrations of the Huttonian Theory*, 445–57 (quotation, 446). On the role of Hutton's theories in establishing what later became known as uniformitarianism, see Simpson, "Uniformitarianism," 45–48; Baker, "Catastrophism and Uniformitarianism," 173–75.
8. Buch, *Travels*, 386–87. On Jameson's critical stance, see Martin, "Translation, Annotation and Knowledge-Making."
9. Buch, *Travels*, 387. For Wrede's work, see Wrede, *Geognostische Untersuchungen*.
10. On the deep motives for resisting Hutton's theory of earth formation, see Eyles, "Hutton, James." On the place of the biblical flood in early modern science, see Barnett, "Theology of Climate Change." On the work of the Swedish navigation service, see Bruncrona and Hällström, "Beobachtungen und Angaben."
11. Lyell, *Principles of Geology*, 1st ed.
12. See Greenough, "Address Delivered at the Anniversary Meeting." For context, see Seibold and Seibold, "Vereisung und Meeresspiegel," 404–5; Rudwick, *Worlds before Adam*, 297–300.
13. On the eighteenth-century debate on sea-level variability, see Hardenberg, "Making a Stable Sea," 134. For an early mention of mean tide as a reference for barometric measurements of height (albeit with no detail on how mean tide level is determined), see Juan and de Ulloa, *Voyage historique de l'Amerique meridionale*, 2:106. On the liberal use of the concept "mean sea level" in the early nineteenth century, see Andrew Scott Waugh, quoted in Phillimore, *Historical Records of the Survey of India*, 5:70. On the preeminence of low water among British surveyors, see Bevan to Wollaston, July 22, 1822. For an account of the selection of the new Irish datum, see Ravenstein, "On Bathy-Hypsographical Maps."
14. On Lalande's rebuttal of the use of mean tide level, see Lalande, *Traité du flux et du reflux*, 113–15. On Laplace's work on the abstraction of the sea's equilibrium, see Laplace, *Traité de mécanique céleste*, 250, 288.
15. On the Brest data series and measurement sessions in the port, see, respectively, Wöppelmann et al., "Tide Gauge Datum Continuity"; Pouvreau, "Trois cents ans de mesures marégraphiques," 74–83. Lalande's attempts to relaunch sea-level gaugings in Brest, and Laplace's correspondence with the maritime prefect and his 1803 report, are mentioned in Pouvreau, "Trois cents ans de mesures marégraphiques," 132–33. For more on the commission to improve tidal observations, see Deparis, Legros, and Souchay, "Investigations of Tides," 78.
16. On the post-Napoleonic mapping effort, see Beautemps-Beaupré and Daussy, *Exposé des travaux*, 1–2, 8. On early state funding of science and the assertion of maritime power, see Burstyn, "Seafaring"; Reidy and Rozwadowski, "Spaces in Between."
17. Daussy, "Mémoire sur les marées," 565. For context, see "Influence de la pression atmosphérique," 136. All translations are mine, if not otherwise attributed.

18. Corabœuf, "Mémoire sur les opérations géodésiques," 45; Corabœuf, "Appendice," 110–124.
19. See Mudge, *Trigonometrical Survey of England and Wales*. On the creation of a sea-level reference standard in Greenwich, see Bevan to Admiralty Office, June 21, 1823.
20. Lloyd, "Account of Operations."
21. Cartwright, *Tides*, 93.
22. On the manual recording of tide data, see Airy to Colby, May 16, 1842; Beautemps-Beaupré and Daussy, *Exposé des travaux*, 9. Airy's contribution to defining mean sea level will be discussed in more detail toward the end of this chapter. On the limits of fieldwork, see Vetter, "Lay Observers."
23. The first description of a working automated gauge is given, anonymously, in *The Nautical Magazine* as "The Tide Gauge at Sheerness." I am grateful to Giulia Danti for sharing with me her work on the history of the Sheerness tide gauge, completed as part of a research internship at the National Maritime Museum in Greenwich..
24. On Young, see Cartwright, *Tides*, 89–90. The description of Palmer's device, which preceded that of the Sheerness gauge by a year, can be found in Palmer and Lubbock, "Description of a Graphical Registrer of Tides and Winds." For more context on Palmer's work, see Matthäus, "On the History of Recording Tide Gauges," 27; Reidy, "Gauging Science and Technology," 11–17. While Reidy claims Palmer's device was never built, I'm less sure. Palmer states in the article presenting the device that he'd had a prototype built. I agree with Reidy, in any case, that it was never installed.
25. On the first wave of automated tide gauges, see Deacon, *Scientists and the Sea*, 257; Matthäus, "On the History of Recording Tide Gauges"; Cliffe, *Book of South Wales*, 28. The first volume of the Admiralty's Tide Tables appeared as *Observations of the Tides*; for context, see Hughes and Wall, "Admiralty Tidal Predictions of 1833." Lenz, "Beschreibung eines sich selbst registrirenden Fluthmessers," describes the tide gauge installed by Russian officers in Alaska.
26. See Deacon, *Scientists and the Sea*, 257–66. BAAS financial support for research on the level of the sea is acknowledged in Whewell, "Account of a Level Line," 2. On the early history of the BAAS, see Morrell and Thackray, *Gentlemen of Science*. On the importance of tide science in Great Britain, see Reidy, *Tides of History*, 122–97. See also Lubbock to Whewell, 1831, in which Lubbock deems Whewell "so much better than [himself]" for the task.
27. Whewell, "Account of a Level Line," 2. On Denham's work, see Denham, "On the Survey of the Mersey and the Dee."
28. On the comparative survey through Somerset and Devon, see Whewell, "Account of a Level Line," 1, 3–7.
29. See Whewell, "Researches on the Tides, Seventh Series," 84. On Whewell's work on tides and his emphasis on spatial comparisons, see Reidy, *Tides of History*, 157–97.
30. Whewell, "Researches on the Tides, Tenth Series," 152–53, 157. On Dessiou's role at the Admiralty, see Hughes and Wall, "Admiralty Tidal Predictions of 1833," 206–7.
31. On Whewell's relationship with his collaborators, see Reidy, *Tides of History*, 270. On Walker's understanding of himself as more than a data collector, see Walker, "Observations on the Tides," 245–251 and 267–273, especially 269–70.

32. Airy, "On the Laws of the Tides on the Coasts of Ireland"; Cartwright, *Tides*, 117.
33. See Woodworth, "Differences between Mean Tide Level and Mean Sea Level." For Airy's remark, see Airy, "On the Laws of Individual Tides at Southampton and at Ipswich," 51.
34. On the importance of standardization to state-building, see Scott, *Seeing Like a State*. See also Sobel, *Longitude*; Dunn and Higgitt, *Finding Longitude*; Bartky, *One Time Fits All*; and Ogle, *Global Transformation of Time*. On the broader modern trend toward standardization as a means of reaching objectivity, see Porter, *Trust in Numbers*. On the infrastructure state, see Guldi, *Roads to Power*.

CHAPTER TWO

1. Beck, "Rechenfehler beim Bau der Hochrheinbrücke."
2. On the use by Belgian state agencies of distinct vertical datums and the later adoption of mean sea level, see Bruhns and Hirsch, *Verhandlungen der . . . Permanenten Commission*, 1877, 79–80; Hirsch, *Verhandlungen der . . . Permanenten Commission*, 1891, 33.
3. For the global spread of automated tide gauges, see Matthäus, "On the History of Recording Tide Gauges." On the San Francisco gauge, see Theberge, "150 Years of Tides."
4. For Humboldt's talk at the academy of sciences, see "Discours prononcé par M. Alexandre de Humboldt," esp. 100. On his use of isometric lines as tools of scientific inquiry, see Dettelbach, "Humboldtian Science," 295–99. On contouring becoming, since the early eighteenth century, the standard method to represent elevations on maps, see Skelton, "Cartography," 612–14. On the later development of Humboldt's ideas, see Humboldt, *Asie centrale*, 288–90. On the Baku marker, see Lenz, "Ueber die Veränderung der Höhe." On Humboldt's Russian expedition and stay in St. Petersburg, see Wulf, *Invention of Nature*, chap. 16.
5. See Ross, *Voyage of Discovery and Research*, 22–24, 317–19. On the Port Arthur tide gauge and what level it actually records, see Hunter, Coleman, and Pugh, "Sea Level at Port Arthur."
6. See "Letter from the Baron Alexander von Humboldt to the Earl of Minto." Others, like the eighteenth-century Bolognese scholar Eustachio Manfredi, preceded Humboldt in lamenting the lack of data needed for long-term diachronic analysis. See Hardenberg, "Making a Stable Sea," 137.
7. The phrase "infrastructure state" is adopted from Guldi, *Roads to Power*, where it refers specifically to the state's role in developing the British road network between 1726 and 1848. On the state's role in public works in, respectively, France, Great Britain, and colonial India, see Geiger, "Planning the French Canals"; Schwartz, Gregory, and Thévenin, "Spatial History"; Ramesh and Raveendranathan, "Infrastructure and Public Works." On the role of colonial bureaucracy in shaping nineteenth-century scientific observations, see Benson, *Surroundings*, 62.
8. See Humboldt, "On the Comparative Level of Lakes and Seas," 335.

9. On Juan and de Ulloa's travels and measurements, see Juan and de Ulloa, *Voyage historique de l'Amerique meridionale*, 2:97; Humboldt, *Essai politique*, 1:239–40. On Walker's stance, see Deacon, *Scientists and the Sea*, 265. On the idea's diffusion among seafarers, see Argonaut, "Permanent Difference of Level," 311, 313, 384. On the origins of the *Nautical Magazine*, see, alas briefly, "Centenary of the Nautical Magazine."
10. Humboldt, *Essai politique*, 1:243.
11. Humboldt, *Relation historique*, 3:556.
12. Lloyd, "Account of Levellings," 63.
13. Lloyd, "Account of Levellings," 63; Arago, "Notices scientifiques," 319; Humboldt, "On the Comparative Level of Lakes and Seas," 332.
14. Rabino, "Statistical Story," 499; Humboldt, *Asie centrale*, 2:327; Argonaut, "Permanent Difference of Level," 385–86. On premodern and early modern theories suggesting a difference in elevation between the Mediterranean and the Red Sea, see Goby, "Histoire des nivellements," 107–14; Wilson, *Suez Canal*, 1–6. On the Canal of the Pharaohs, see Redmount, "Wadi Tumilat." On how historical accounts and geological findings affected the development of public works in the nineteenth century, see Chakrabarti, *Inscriptions of Nature*, 1–54. On tidal measurements in the Suez region, see Goby, "Marées de la Mer Rouge," 17–23.
15. See Goby, "Histoire des nivellements," 114–19. For a sense of the extent to which difficult conditions forced colonial surveyors to improvise, see Schaffer, "Oriental Metrology," 181.
16. See Goby, "Histoire des nivellements," 127–31; Le Père, "Mémoire sur la communication," 21–31. For an example of Arago's unquestioning adoption of Le Père's results, see Arago, "Sur les phénomènes de la mer," 586. On the role of trust in making Le Père's results more reliable, see Montel, "Établir la vérité scientifique," 89–90, 95. On proposals to overcome or exploit the measured height difference, see Chevalier, "L'isthme de Panama," 67–68; Wilson, *Suez Canal*, 10. On early nineteenth-century objections to the construction of a canal, see Rabino, "Statistical Story," 498–99.
17. For the details of Bourdaloüe's Suez survey, see Goby, "Histoire des nivellements," 137–43. On the consortium that sponsored it, see Wilson, *Suez Canal*, 9–10. On Bourdaloüe's vertical datum, see Bourdaloüe, "Notice sur le nivellement de l'isthme de Suez," 7–8; Bateman, "Some Account of the Suez Canal," 133. On the elevation difference determined by Bourdaloüe in relation to the possible range of error and the first full series of tidal measurements, see Lesseps, *Percement de l'Isthme de Suez*, 239–41.
18. On the issue of precision in Bourdaloüe's work, see Montel, "Établir la vérité scientifique," 94–100. On other early nineteenth-century surveys, see Goby, "Histoire des nivellements," 134–37.
19. On the various reference levels used in France prior to 1860 and the role of infrastructural development in pushing for their standardization, see Lallemand, "L'unification des altitudes," 931–32.
20. On how the new French vertical datum was selected, see Coulomb, *Le marégraphe de Marseille*, 55–61; Ministère . . . des Travaux publics, *Nivellement de la France*,

V–VI; Bourdiol, "Importance d'un nivellement général de la France," 187; Chaumont, "Ministére des Travaux publics," 6. The vertical datum for Corsica, separated from the mainland by a significant water expanse, was defined as mean sea level in Ajaccio. Its first known assessment was, however, done only between 1912 and 1937. See IGN, "Les réseaux de nivellement français"; Cariou, "Où se trouve le niveau de la mer?"

21. Woodworth, "Study of Changes," 21–22; Bradshaw et al., "Century of Sea Level Measurements at Newlyn," 116. On Airy's work for the Irish survey, see chapter 1.
22. On the selection of Britain's first national datum, see Woodworth, "Sea Level Change in Great Britain," 3; James, *Abstracts of the Principal Lines of Spirit Levelling*, v–vi; Woodworth, "Study of Changes," 22. The gauge at George's Pier in Liverpool exploited the flotation of the pier's landing stage rather than a float in a stilling well; thus, movements caused by daily use could introduce a degree of variability. Moreover, low-water data could be lost since the stage could ground during low spring tides. On the social nature of reference points, datums, and baselines, see Ureta, Lekan, and Hardenberg, "Baselining Nature."
23. See Phillimore, *Historical Records of the Survey of India*, 2:17, 257–59; Lambton, "Account of the Measurement of an Arc," 136–93. For the specific case of Bengal, see Phillimore, *Historical Records of the Survey of India*, 1:347, 3:14.
24. See Malcolm, *Memoir of Central India*, 2:348; Phillimore, *Historical Records of the Survey of India*, 3:38, 203–5; Hodgson and Herbert, "Account of Trigonometrical and Astronomical Operations," 209, 319–20. See also Fleetwood, "No Former Travellers," for a detailed analysis of the use of instruments to assess elevation in the Himalayas, esp. 15–18 on the risk of breaking a barometer on the field, the consequences of such accidents, and the limitations of boiling-point thermometers.
25. Letter from Court of Directors, February 24, 1834, quoted in Phillimore, *Historical Records of the Survey of India*, 4:119. On the conflictual relationship between natural philosophers and paid laborers in early modern science, see Shapin and Schaffer, *Leviathan and the Air-Pump*, 125–39; Shapin, "Invisible Technician." For an example of the condescending attitude of some nineteenth-century scientists toward the manual workers crucial to their work, see Rozwadowski, *Fathoming the Ocean*, 194. For a review of the historiographical debate about professionalization in science, see Lucier, "Professional and the Scientist."
26. See Phillimore, *Historical Records of the Survey of India*, 5:69–70. On the changing nature of the coast by Calcutta, see Bhattacharyya, *Empire and Ecology*, 45–76.
27. On the first automated tide gauges on the Indian subcontinent and the quality of their measurements, see Phillimore, *Historical Records of the Survey of India*, 5:69–72. On the history of Colaba Observatory, see Charles Chambers, *Meteorology of the Bombay Presidency*, 4–5. For Waugh's quotes, see Phillimore, *Historical Records of the Survey of India*, 5:23 (1845), 4:94 (1853).
28. Phillimore, *Historical Records of the Survey of India*, 5:66–67, 73–75.
29. Waugh to Nasmyth, June 25, 1854, quoted in Phillimore, *Historical Records of the Survey of India*, 5:73.
30. Phillimore, *Historical Records of the Survey of India*, 5:68.

CHAPTER THREE

1. See Bouchayer, *Marseille*, 37–38. For a brief biographic account, see Blanchard, "M. Auguste Bouchayer."
2. On Bouchayer's interest in sea-level variability and some biographical details on Le Doyen, see Bouchayer, *Marseille*, 11–15, 30–31 (quotation, 30). I will not list here the many reviews of Bouchayer's book; a title search on Gallica, the digital library of the Bibliothéque national de France (https://gallica.bnf.fr/) will give an impression of its circulation in France in the early 1930s.
3. Lallemand, "Note sur les travaux," 932–33; Heyde, *Die Höhennullpunkte*, 13–19.
4. On the Belgian vertical datum of 1879 and the first reference points used in Prussia, see Spata, "Historische Pegel und Bezugshöhen in Europa," 381, 383. On the Swinemünde tide gauge, see Torge, *Geschichte der Geodäsie in Deutschland*, 225–26; Baeyer, *Nivellement zwischen Swinemünde und Berlin*, 81. The Permanent Service for the Mean Sea Level website gives 1811 as the start date for sea-level gaugings at Swinemünde, but I could locate no source confirming the reliability of data from before 1826: https://www.psmsl.org/data/obtaining/stations/2.php. On the idea of a sloping Baltic Sea, see "Das Niveau der Ostsee," 229–30. On perceived limitations of the existing measuring network, see Meyer et al., *Jahresbericht der Commission*, 4.
5. "Der Normal-Höhenpunkt," 3.
6. See "Der Normal-Höhenpunkt," 4. On the history of attempts to determine the level of the sea offshore, see Hardenberg, "Measuring Zero at Sea."
7. "Der Normal-Höhenpunkt," 1–2.
8. On the realization that the Amsterdam datum and the *Normalnull* differed, see Heyde, *Die Höhennullpunkte*, 6. On the swift adoption of the new datum throughout the Reich, see "Der Normal-Höhenpunkt," 12–16. Each German state used to have its own datum: Bavaria, for instance, adopted the mean level of the Adriatic Sea in Venice, while surveys in Baden referred to the floor of the Strasbourg cathedral, the elevation of which the French had determined in relation to the mean level of the Mediterranean in Marseille. Spata, "Historische Pegel und Bezugshöhen in Europa," 383.
9. On the history of German colonialism, see Conrad, *Deutsche Kolonialgeschichte*.
10. Moser, "Untersuchungen zur Kartographiegeschichte von Namibia," 111.
11. Hafeneder, "Deutsche Kolonialkartographie 1884–1919," 138; Conrad, *Deutsche Kolonialgeschichte*, 33–34.
12. On German cartography in the Pacific, see Hafeneder, "Deutsche Kolonialkartographie 1884–1919," 121–33. On the history of cartography in New Guinea, see Linke, "Influence of German Surveying," 17.
13. See Hafeneder, "Deutsche Kolonialkartographie 1884–1919," 158; Kortum, "Naval Observatory in Tsingtau," 259–62. Because of the length of the available data series, mean sea level at the Qingdao tide gauge was selected in 1957 as China's basic height datum; see Wu, Zeng, and Ming, "Analyzing the Long-Term Changes," 1342–43.
14. Baeyer, *Zur Entstehungsgeschichte der europäischen Gradmessung*, 2–5; Ohnesorge, "How Incoherent Measurement Succeeds," 252–58, 255; Bruhns, Förster, and Hirsch, *Bericht über die Verhandlungen der . . . allgemeinen Conferenz*, 142.

15. To avoid confusion caused by the very similar English names of the geodesy associations before and after World War I, I will follow the example of Walter D. Lambert—an early historian of the associations and IAG's president in the late 1940s—and use the German denomination Internationale Erdmessung throughout the text. Lambert, "International Geodetic Association," 299.
16. On the association's founding, see Baeyer, *Zur Entstehungsgeschichte der europäischen Gradmessung*. Baeyer's friendship with Bessel is recorded in Peirce to Mills, November 2, 1877. On the political implications of terming the association "Central European," see Ogle, *Global Transformation of Time*, 28. On the history of the commission and its further internationalization, see Völter, *Geschichte und Bedeutung*, 7–11. On Great Britain's delay in joining the association, see Darwin, "British Empire," 511. On the conflictual relationship of British imperialism with scientific internationalism, see Mahony, "For an Empire of 'All Types of Climate.'" On the history of IUGG, see also chapter 5.
17. See Baeyer, *Zur Entstehungsgeschichte der europäischen Gradmessung*, 13; Förster, *Verhandlungen der ersten allgemeinen Conferenz*, 11; Tirkot, "Bibliothek des Geo-Forschungszentrum (GFZ)"; Torge, "From a Regional Project to an International Organization," 8–9. Reports of the geodetic associations active between 1862 and 1912, digitized in cooperation with GeoForschungsZentrum Potsdam as part of the Measuring the Earth project, are available through the digital library of the Max Planck Institute for the History of Science (https://dlc.mpg.de/partner/mpiwg/) under "Generalberichte und Verhandlungen der allgemeinen Conferenz der Internationalen Erdmessung."
18. Förster, *Verhandlungen der ersten allgemeinen Conferenz*, 27–29.
19. Bruhns, Förster, and Hirsch, *Bericht über die Verhandlungen der . . . allgemeinen Conferenz*, 148.
20. On Schiavoni's request, see Bruhns and Hirsch, *Bericht über die Verhandlungen der . . . dritten allgemeinen Conferenz*, 76–77. Baeyer's daughter's commentary is recorded in a letter Peirce sent to his mother from his third trip to Europe; Peirce to Mills, November 2, 1877.
21. Bruhns and Hirsch, *Verhandlungen der . . . Permanenten Commission* (1876), 209.
22. For overviews of the European search for a common zero of elevation, see Lallemand, "Rapport présenté au nom de la Commision"; Heyde, *Die Höhennullpunkte*, 4. For Ibáñez's reports, see Ibáñez, "Rapport sur l'état des travaux" (1881; 1884; 1890). For Peirce's remark, see Peirce to Mills, November 2, 1877. For Hirsch's comments and sources, see Hirsch, "Rapport sur l'état actuel des travaux," 12; Hirsch and Oppolzer, *Verhandlungen der . . . Permanenten Commission*, 1883, 51–52. Baeyer's theory is reported in Bruhns and Hirsch, *Verhandlungen der . . . Permanenten Commission* (1876), 209–10.
23. Lallemand, "Note sur le principe fondamental," 133–35; Hirsch, *Verhandlungen der . . . allgemeinen Conferenz* (1887), 43–45, 49.
24. Hirsch, *Verhandlungen der . . . Permanenten Commission* (1889), 39–42.
25. Lallemand, "Note sur les travaux"; Hirsch, *Verhandlungen der . . . allgemeinen Conferenz* (1890), 34.

26. Hirsch, *Verhandlungen der . . . Permanenten Commission* (1891), 78; Lallemand, "Note sur l'unification."
27. See Lallemand, "L'unification des altitudes," 938–39. On precision and accuracy in nineteenth-century geodesy, see Ohnesorge, "Promises and Pitfalls of Precision."
28. Helmert, "Le zéro des altitudes"; Hirsch, *Verhandlungen der . . . Permanenten Commission* (1892), 44–45.
29. Börsch, "Vergleichung der Mittelwasser."
30. For Ferrero's comments and the ensuing debate, see Hirsch, *Verhandlungen der . . . allgemeinen Conferenz* (1893), 113–17. On the diplomacy of governmental internationalism, see also Herren, "Governmental Internationalism"; Geyer, "One Language for the World."
31. The quote is from Hirsch, *Verhandlungen der . . . allgemeinen Conferenz* (1893), 115. The marker of the Swiss datum had been set, in 1820, on an erratic block in the port of Geneva, its elevation determined in respect to the mean level of the ocean at Noirmoutier, on the French Atlantic coast, as measured by the French geodesist Jean-Baptiste Corabœuf. See Heyde, *Die Höhennullpunkte*, 18–19.
32. Hirsch, *Verhandlungen der . . . allgemeinen Conferenz* (1893), 116–17.
33. Lallemand, "Rapport présenté au nom de la Commision," 132–34.
34. See, for instance, Löschner, "Zur Frage der Vereinheitlichung," 101.
35. On the Austrian case, see Löschner, "Zur Frage der Vereinheitlichung," 90.
36. On abstraction in the choice of vertical datums, see Hardenberg, "Measuring Zero at Sea."
37. On the political nature of standards, see Wise, "Precision," 92–100.

CHAPTER FOUR

1. Suess, *Erinnerungen*, 138–39. An excellent, if brief, history of eustatic theories is given by Dott, *Eustasy*. On plate tectonics, see also Trewick, "Plate Tectonics in Biogeography."
2. See Baker, "Catastrophism and Uniformitarianism." On the Holocene as a "long summer," see Flannery, *Weather Makers*, chaps. 6 and 7. For a summation of the debate about the Anthropocene, see Bińczyk, "Most Unique Discussion." On the revival of catastrophist thinking in debates on the Anthropocene, see Mauelshagen, "Climate Change, Decline and Societal Collapse." See also Chakrabarty, *Climate of History in a Planetary Age*, 1–14. For an introduction to the ice age controversy, see Davies, *Earth in Decay*, 263–316.
3. Maclaren, " Glacial Theory of Prof. Agassiz," 365; Robert, "Recueil d'observations," 265–67.
4. See Adhémar, *Révolutions de la mer*. For Lyell's map of a flooded Europe, see Lyell, *Principles of geology*, 9th ed., 121–24. On coeval criticisms of Adhémar's theories, see Krüger, *Discovering the Ice Ages*, 405–6; Imbrie and Imbrie, *Ice Ages*, 75. On their positive impact on the debate, see Wood, "Climate Delusion," 11. As an example of the kind of works published, also by amateurs, in response to Adhémar's theory, see Bruchhausen, *Die periodisch wiederkehrenden Eiszeiten und Sindfluten*.

5. Adhémar, *Révolutions de la mer*, 21–30.
6. See Chambers, *Ancient Sea-Margins*, 18. For context, see also Hestmark, "Tracings of the North of Europe."
7. Darwin, *Structure and Distribution of Coral Reefs*; Chambers, *Ancient Sea-Margins*, 18–22.
8. Tylor, "On Changes of the Sea-Level," 262.
9. Tylor, "On Changes of the Sea-Level," 264. On the eighteenth-century debate about silt as a cause for global sea-level rise, see Hardenberg, "Making a Stable Sea."
10. Tylor, "On the Formation of Deltas," 392–99, 485–501; "Proceedings of the Geological Society," 1–12. For context, see also Darwin, *Structure and Distribution of Coral Reefs*; Droxler and Jorry, "Origin of Modern Atolls."
11. Whittlesey, "Depression of the Ocean." Compared with more recent estimates, Whittlesey's calculation underestimates the potential impact of the continued melting of the planet's ice caps, mostly due to its extremely conservative assessment of the thickness of the Antarctic ice cover. His ballpark figures for the possible level of the sea at the height of the last ice age concur instead with present day estimates: late twentieth-century estimates of the level of the sea seventeen thousand years ago indicate that it was about 120 meters below the current level. See Davies, "Calculating Glacier Ice Volumes"; O'Hara, *Brief History of Geology*, 189.
12. See Wolf, "Changing Role of the Lithosphere," 96–97. The theory was confirmed toward the end of the nineteenth century by Gerard De Geer's work on old Swedish shorelines; see De Geer, "Om Skandinaviens nivåförändringar." As part of a more general standardization of naming practices, the theory was later given the now more commonly accepted term "glacial isostasy" by the American geologist Clarence Dutton. See Dutton, "On Some of the Greater Problems of Physical Geology," 51–64.
13. Croll, "On the Physical Cause of the Submergence of the Land," 270–71. That Croll was unaware of Adhémar's work when he first wrote on the issue is mentioned in the footnote on page 368 of Croll, *Climate and Time*. On Croll's uniformitarian credentials and his contribution to the theory of multiple glaciations, see Fleming, "Pathological History of Weather," 44–46. On the public reception of Croll's theory, see Irons, *Autobiographical Sketch of James Croll*, 493–94. The mechanics of Croll's take differ significantly from current interpretations, but recent research in geophysics shows how Greenland's ice sheet indeed attracts water, elevating the current level of the North Atlantic in respect to other seas. When the ice sheets melt, this gravitational effect will disappear as well; water will flow away from the island and, in a sort of rebound effect, contribute to an even greater rise in sea level farther away. See Hsu and Velicogna, "Detection of Sea Level Fingerprints."
14. For criticisms of Croll's view, see Wood, "Glacial Submergence"; Fleming, "Charles Lyell and Climatic Change," 167. For Croll's reaction, see Croll, "On the Change in the Obliquity of the Ecliptic," 191. On the genealogy of the idea of ancient beaches, see Fairbridge, "Raised Beach"; Croll, *Climate and Time*, 406–7.
15. See Thomson, "Polar Ice-Caps and Their Influence," 326, 327. The note was first published in Croll, *Climate and Time*, 372–74. On the success of Croll's book, see Fleming, "Pathological History of Weather," 45.

16. Trautschold, *Ueber säkulare Hebungen und Senkungen*; Trautschold, *Sur l'invariabilité du niveau des mers*.
17. Shaler, "Notes on Some of the Phenomena," 288, 292.
18. Penck, *Schwankungen des Meeresspiegels*.
19. See Listing, *Ueber unsere jetzige Kenntniss*, 8–10; Penck, *Schwankungen des Meeresspiegels*, 4–5; Hann, "Ueber gewisse beträchtliche Unregelmässigkeiten," 563. For a discussion of attempts made at the time to avoid the gravitational distortions caused by continental masses by determining sea level at a distance from the coast, see Hardenberg, "Measuring Zero at Sea."
20. Hirsch, *Verhandlungen der . . . Permanenten Commission* (1892), 148; Hirsch, *Verhandlungen der . . . Permanenten Commission* (1888), 62–63.
21. Suess, *Das Antlitz der Erde*, 2:680–88.
22. Suess, *Das Antlitz der Erde*, 2:693–95, 702.
23. Thompson, "On Mean Sea Level." Regarding measurements on the North American coast, see Johnson and Winter, "Sea-Level Surfaces," 475.
24. Quoted in Dawson, *Tide Levels and Datum Planes*, 11–12.
25. Dawson, *Tide Levels and Datum Planes*, 12.
26. Davidson, "États-Unis," 4; Rappleye, "Sea-Level Datum of 1929"; *Annual Report of the Superintendent*, 47–49. On New York City, see Koop, *Precise Leveling in New York City*, 71–78. For context on standardization in the United States, see Jones, *Use of Mean Sea Level*.
27. *Annual Report of the Superintendent*, 49.
28. "Standards and Procedures," chap. 2, 4–5; Lachapelle, "Status of the Redefinition of the Vertical Reference System," 610.
29. Royal Commission on Coast Erosion and Afforestation, *Third (and Final) Report*, 3–6; Reid, *Submerged Forests*. The idea that the submersion of ancient forests was primarily due to subsidence was proposed as early as 1799 and then used by Playfair to support the idea of a stable sea. Correa de Serra, "On a Submarine Forest," 152–53; Playfair, *Illustrations of the Huttonian Theory*, 452–53.
30. On the process that led to revising the British datum, see Henrici, "Mean Sea-Level." On how the Ordnance Survey chose Newlyn as its reference station, see Bradshaw et al., "Century of Sea Level Measurements at Newlyn"; Close, *Second Geodetic Levelling*. For the quote, see Johnson, "Is the Atlantic Coast Sinking?" 136.
31. Suess, "Zur Deutung der Vertikalbewegungen" (1920); Johnson, *Fixité de la Côte Atlantique*, esp. 212. A brief remark on Johnson's take on the variability of the sea is given by Fairbridge, "Eustatic Changes in Sea Level," 99.
32. Spencer, "Glacial Control Hypothesis"; Daly, "Glacial-Control Theory"; Daly, "Recent Worldwide Sinking," 248. On the long-term impact of Daly's theory, see Bloch, "Hypothesis for the Change of Ocean Levels," 127.
33. On the 1920s debate on the role of eustasy in explaining sea-level changes, see Suess, "Zur Deutung der Vertikalbewegungen" (1921); Johnson and Winter, "Sea-Level Surfaces."
34. Penck, "Eustatische Bewegungen des Meeresspiegels," 329–35; Ramsay, "Changes of Sea-Level."

35. "Climate Changes and the Last Glacial Period," 9.
36. See Baulig, *Changing Sea Level*, esp. 2 for the quote. For an impression of how glacio-eustatism quickly became the theory of reference to explain postglacial sea-level changes, see Godwin, "Coastal Peat Beds," 208.
37. Baulig, *Changing Sea Level*, 2.

CHAPTER FIVE

1. Witting, "Mean Sea Level," 41, 43. For the history of IUGG, see Ismail-Zadeh and Joselyn, "IUGG." IUGG was set up as a federation of specialty groups and a member of the International Research Council (IRC), created jointly by the scientific academies of the allied nations in 1919. See Greenaway, *Science International*, 18–32; Cock, "Chauvinism and Internationalism in Science." See also Herren, "Governmental Internationalism." Regarding efforts made by the IAPO Tidal Committee to collect information about existing tide gauges and to standardize the data-gathering processes, see bulletins of the Section d'oceanographie physique: *Réunion plénière de Paris*, 18; *Réunion plénière: Rome*, 52; *Réunion plénière . . . (Madrid)*, 30; *Réunion plénière . . . (Prague)*, 27. On the formation and work of the mean sea level committee, see Cartwright, "Historical Development of Tidal Science," 246; Association d'Océanographie Physique, *General Assembly at Edinburgh*; Association d'Océanographie Physique, *Monthly and Annual Mean Heights of Sea-Level, Up to and Including the Year 1936*; Association d'Océanographie Physique, *Monthly and Annual Mean Heights of Sea-Level 1937 to 1946*; Association d'Océanographie Physique, *Monthly and Annual Mean Heights of Sea-Level 1947 to 1951*. On the genealogical connection between the IAPO mean sea level committee and PSMSL, see Rossiter, "Les travaux."
2. See Warde, Robin, and Sörlin, *Environment*.
3. Lane et al., "Round Table Discussion," 139–40. Sigurdur Thorarinsson is an Icelandic name. Thorarinsson is thus technically not a surname but a patronymic. According to Icelandic naming conventions, the author should be cited either by his full name or as Sigurdur. To avoid confusion here, in a nonspecialist publication, I nonetheless follow the common English practice and use Thorarinsson as if it were a surname.
4. Thorarinsson, "Present Glacier Shrinkage," 134, 151.
5. Thorarinsson, "Present Glacier Shrinkage," 150–52. For an example of Thorarinsson's earlier works on Vatnajökull, see Ahlmann and Thorarinsson, "Vatnajökull." For context on Ahlmann and his polar warming theory, see Sörlin, "Global Warming That Did Not Happen," 97–100; Sörlin, "Anxieties of a Science Diplomat." On the wider debate in the 1920s and 1930s about Scandinavia's postglacial rebound, see Ramsay, *On Relations between Crustal Movements and Variations of Sea-Level*; Post, "Gothiglacial Transgression." For a broader history of the issues at stake, see Wolf, "Changing Role of the Lithosphere."
6. Gutenberg, "Changes in Sea Level," 730.
7. Kuenen's work of 1945 and his later analyses are reported in Kuenen, *Marine Geology*, 533–35 (quotation, 535).
8. Bloch, *Historian's Craft*, 24.

9. See Johnson, "Is the Atlantic Coast Sinking?"; Marmer, "Is the Atlantic Coast Sinking?" 655–57.
10. Association d'Océanographie Physique, *Secular Variation of Sea-Level*, 3.
11. Landes's provocation and a critical response are given in Wheeler, "Sanctity of Sea Level," 1325. See also Fairbridge, "Eustatic Changes in Sea Level," 99. For Egedal's take on the need to make sea-level research global, see Association d'Océanographie Physique, *Secular Variation of Sea-Level*, 6.
12. Munk and Revelle, "Sea Level and the Rotation of the Earth," 833; Munk and Revelle, "On the Geophysical Interpretation of Irregularities," 335.
13. On Teller and his predictions, see Franta, "On Its 100th Birthday."
14. Teller, "Energy Patterns of the Future," 70.
15. See Ballard, *Drowned World*; Tiny Tim, *Other Side*. For context on Ballard's work, see Haith, "Sea-Level Rise."
16. See Bloch, "Hypothesis for the Change of Ocean Levels," 128, 133–34.
17. For examples of work done on the topic since the mid-1950s, see Shepard and Suess, "Rate of Postglacial Rise," 1083; Godwin, Suggate, and Willis, "Radiocarbon Dating"; Fairbridge, "Dating the Latest Movements"; Fairbridge, "Eustatic Changes in Sea Level"; Jelgersma, *Holocene Sea Level Changes*.
18. See Fairbridge, "Changing Level of the Sea," 79; Donn and Shaw, "Sea Level and Climate of the Past Century." On the development of the idea that humans could cause climate change and, consequently, sea-level rise, see Sörlin, "Global Warming That Did Not Happen," 95.
19. Endeavors to decouple geodesy from the vagaries of local sea-level gaugings were made long before the satellite era, as early as the nineteenth century. On the history of these attempts, see Hardenberg, "Measuring Zero at Sea." On the idea of a neutral observation point, see Hardenberg and Mahony, "Up, Down, Round and Round," 606.
20. On remote sensing, see Höhler and Wormbs, "Remote Sensing." About the distinction between the *global* as an anthropocentric dimension and the *planetary* as one that allows a decentering of the human, see Chakrabarty, *Climate of History in a Planetary Age*, esp. 1–4.
21. For context, see Polezhayev, "Use of Artificial Earth Satellites" (Rosenberg's quote is on page 2); Lambeck and Coleman, "Earth's Shape and Gravity Field." Two exemplary articles published in 1958 are Cook, "Determination of the Earth's Gravitational Potential"; Jacchia, "Earth's Gravitational Potential," 19. On and by Kaula, see, respectively, Turcotte, *William M. Kaula 1926–2000*; Kaula, *Theory of Satellite Geodesy*.
22. On the match between geophysicists and remote-sensing engineers, see Wilson, Lindstrom, and Apel, "Satellite Oceanography"; Ewing, *Oceanography from Space*, vii–ix. On the role of satellites in overcoming the dearth and biased distribution of data points, see Munk, "Oceanography before, and after, the Advent of Satellites," 1–3; Adler, "Ship as Laboratory," 353. For context on the IGY's role in fostering oceanographic sea-level research, see Hamblin, *Oceanographers and the Cold War*, 74–76.
23. Malakhov, "Marine Geodesy"; Kutazov, "Scientific Problems of Geodesy and Cartography."

24. Lambeck and Coleman, "Earth's Shape and Gravity Field"; Revelle, "Oceanography from Space"; Yamarone, Resell, and Farless, "TOPEX/Poseidon Mapping."
25. Lehman, "From Ships to Robots"; Tamisiea et al., "Sea Level." On the early history of the Global Level of the Sea Surface (GLOSS) program of the Intergovernmental Oceanographic Commission (IOC), see Pugh, "Improving Sea Level Data," 64–66. For an example of climate skepticism, see Alexander, "No Evidence That Climate Change Is Accelerating Sea Level Rise." A comparison of satellite and tide-gauge data series for 1990–2013 is provided in EPA, *Climate Change Indicators in the United States*, 34.
26. On the radical improvement in the margin of error, see Church et al., "Changes in Sea Level," 663–64; Shum, Ries, and Tapley, "Accuracy and Applications of Satellite Altimetry," 325. On OSTM/Jason-2 and Jason-3, see Dumont et al., *OSTM/Jason-2 Products Handbook*, 8; National Environmental Satellite, Data, and Information Service, "JASON-3 Mission"; EUMETSAT, "Jason-3 Instruments."
27. On errors detected in satellite height measurements, see Tollefson, "Satellite Snafu"; Watson et al., "Unabated Global Mean Sea-Level Rise." On the role of international politics in the history of satellite technology, see Lambright, "Political Construction of Space Satellite Technology." See also Höhler and Wormbs, "Remote Sensing," 280; Lehman, "From Ships to Robots."
28. Broecker, "Climatic Change," 463. On Broecker's pioneering role in the "global warming" debate, see Krajic, "Wallace Broecker."
29. Etkins and Epstein, "Rise of Global Mean Sea Level."
30. For a survey of estimates of recent historical, ongoing, and future global sea-level change, see Warrick and Oerlemans, "Sea Level Rise," 263, 275. The EPA report has been published as Hoffman, Keyes, and Titus, "Projecting Future Sea Level Rise." On its reception, see Sullivan, "So Far, Greenhouse Effect Heats Only Debate." For the comment by Reagan's science advisor, see Shabecoff, "Haste of Global Warming Trend Opposed." For context, see also Edwards, *Vast Machine*, 389.
31. Menefee, "Half Seas Over," 186–87.
32. Rich, "Losing Earth."
33. Menefee, "Half Seas Over," 190, 194, 198–200, 215–16. On the issue of how sea-level variations affect maritime boundaries, see Sammler, "Rising Politics of Sea Level."
34. Warrick and Oerlemans, "Sea Level Rise," 266, 274–77.
35. On NAVD 88 see Sammler, "Rising Politics of Sea Level," 612. On the history of the UELN, see Spata, "Historische Pegel und Bezugshöhen in Europa," 388. For more on the context in which the new North American continental datum is being determined, see chapter 6.

CHAPTER SIX

1. *An Inconvenient Truth*, Documentary (Paramount Classics, 2006).
2. Tingle, "Flood Maps," http://flood.firetree.net/?ll=52.6813,5.5283&zoom=7&m=1. On this tool's limitations, see also Tingle, "More about Flood Maps." Another sea-level mapping tool with similar limitations is "Flood Map," https://www.floodmap.net/. See also Davies, "If All the Ice in Antarctica Were to Melt."

3. Climate Central, "Comparison: Long-Term Sea Level Outcomes." See IPCC, "Global Warming of 1.5°C," and the October 13, 2021, tweet by Pinboard, responding to a map published in the *Boston Globe*: "The unintentional lesson in this map is that for Boston, it doesn't really matter much whether we stop emitting greenhouse gases tomorrow or just relax and keep on going. The 1.5°C of warming responsible for most of the flooding is already baked in" (https://twitter.com/Pinboard/status/1448132568474271745). For the NOAA tool, see National Oceanic and Atmospheric Administration, "Sea Level Rise Viewer," https://coast.noaa.gov/slr/#/layer/slr.
4. On past and modeled rates of increase in global mean sea level, see Sammler, "Rising Politics of Sea Level," 614; Sweet et al., "Sea Level Rise."
5. See Regnauld and Limido, "Coastal Landscape as Part of a Global Ocean," esp. 10–11 on relative sea level; Sörlin and Wormbs, "Environing Technologies"; Wickberg and Gärdebo, "Where Humans and the Planetary Conflate." On sea level and boundaries as legal constructs, see Sammler, "Rising Politics of Sea Level," 605, 612–13; Bhattacharyya, *Empire and Ecology*, 111–39. On legibility, see Scott, *Seeing Like a State*, 11–52.
6. See Houghton et al., "Maritime Boundaries," 814–15.
7. For an introduction to the role of computation in modern climate science, see Edwards, *Vast Machine*, 251–85, 357–96.
8. On the possible use of optical atomic clocks to measure heights, see "What Happens to a Bridge?"; "Wie hoch liegt ein Ort?"
9. On the discrepancy between geodetic heights and practical applications, see Vaníček,"On the Global Vertical Datum," 244; Sammler, "Rising Politics of Sea Level," 612–13.
10. I spoke with Daniel Roman on April 5, 2016; we further discussed the topic in an email exchange on November 30, 2021. For more on the International Height Reference System, defined by the IAG in 2015, see Sánchez et al., "Strategy for the Realisation of the International Height Reference System (IHRS)," 33.
11. On the difficulties of connecting the local and the planetary, see Chakrabarty, *Climate of History in a Planetary Age*, 14, 44.
12. For census data on vulnerable regions, see Cohen, "About 60.2M Live in Areas Most Vulnerable to Hurricanes." On the predicted impact of climate change–induced sea-level rise on US military installations worldwide, see Ghosh, *Nutmeg's Curse*, 125. For an example of a Virginia settlement slowly succumbing to the rising sea, see Johnson, "What the Tiny Tangier Island Teaches Us." On the connection between sea-level rise and hurricane storm surges, see Rahmstorf, "Rising Hazard"; Strauss et al., "Economic Damages from Hurricane Sandy." On the role of race and class in determining the social impact of extreme weather events, see Elliott and Pais, "Race, Class, and Hurricane Katrina"; Faber, "Superstorm Sandy and the Demographics of Flood Risk"; Misra, "Catastrophe for Houston's Most Vulnerable People."
13. On conceptualizing disaster as a process rather than a discrete event, see Knowles, "Slow Disaster in the Anthropocene."
14. Miller, "Heaven or High Water."

15. See Englander, "Does Flood Insurance Cover Rising Seas?"; Miller, "Heaven or High Water"; Ariza, "As Miami Keeps Building, Rising Seas Deepen Its Social Divide"; Cusick, "Miami Is the 'Most Vulnerable' Coastal City."
16. On the Nicobar Islands and the history of colonialism, see Ghosh, *Great Derangement*, 35–37 (quotation, 35). See also "Andaman, Nicobar Islands May Not Be Inhabitable in Future Due to Rise in Sea Level: IPCC." On the (re)appreciation of the sea as a modern phenomenon, see Corbin, *Lure of the Sea*.
17. See Ghosh, *Great Derangement*, 48–53 (quotation, 48). On the reclamation of Mumbai's tidal plains started by the East India Company, see Dossal, *Theatre of Conflict*, 18–19, 39–42.
18. For a case study of how land and water have been made into discrete categories within colonial frameworks, see Bhattacharyya, *Empire and Ecology*. On Bangladesh, see Butzengeiger and Horstmann, *Sea-Level Rise*, 5. On the Netherlands, see Nienhuis, *Environmental History*, 568–73.
19. On Cyclone Bhola and the birth of Bangladesh, see Biswas and Daly, "'Cyclone Not above Politics.'" On South Talpatti/New Moore and "international legal imaginaries," see Bhattacharyya, "A River Is Not a Pendulum," 147–48. See also Schofield, "Holding Back the Waves?" 601. On the "vanishing islands" of the Sundarbans, see "Disappearing World." On environmental migration, see Biasillo, "Historical Tools."
20. On the possible impact of sea-level rise on the coasts of Bangladesh, see Butzengeiger and Horstmann, *Sea-Level Rise*, 6; Huq, Ali, and Rahman, "Sea-Level Rise and Bangladesh." On migration induced by sea-level rise, see Hauer et al., "Sea-Level Rise and Human Migration." On the limits of the country's network of tide gauges, see Khandker, "Mean Sea Level in Bangladesh."
21. On effective sea-level rise, see Pethick and Orford, "Rapid Rise in Effective Sea-Level."
22. See Wanless and Harlem, "Statement on the Evidence." On an earlier use of barnacles and algae to assess and compare sea level, and on the ecology of barnacles, see Kaye, "Upper Limit of Barnacles." On the Great Acceleration, see Steffen et al., "Trajectory of the Anthropocene."
23. "Tuvalu Minister Stands in Sea to Film Cop26 Speech"; "Maldives Sends Climate SOS." Together with eight other countries, Tuvalu has formed the Commission of Small Island States on Climate Change and International Law to pursue enforcement of international law in respect to the existential threat of climate change. Quell, "Island Nations Demand Climate Change Action."
24. See Royal Commission on Coast Erosion and Afforestation, *Third (and Final) Report*. On Doggerland and the continued impact of erosion on England's eastern coast, see Macfarlane, "Silt," 52–54. On the historical reliability of Aboriginal oral histories, see Nunn and Reid, "Aboriginal Memories of Inundation." See also Albrecht, "'Solastalgia': A New Concept"; Albrecht et al., "Solastalgia: The Distress Caused by Environmental Change."
25. On the understanding of Māori and other Indigenous theorists of the land-sea binary, see Sammler, "Kauri and the Whale," 81. On the instability of the geoid, see Tamisiea et al., "Sea Level," 2–3.

REFERENCES

Adhémar, Joseph Alphonse. *Révolutions de la mer*. Paris: Carilian-Goeury et V. Dalmont, 1842.

Adler, Antony. "The Ship as Laboratory: Making Space for Field Science at Sea." *Journal of the History of Biology* 47, no. 3 (2014): 333–62.

Ahlmann, H. W:son, and Sigurdur Thorarinsson. "Vatnajökull: Scientific Results of the Swedish-Icelandic Investigations 1936–37." *Geografiska Annaler* 19 (1937): 146–231.

Airy, George Biddell. Letter to Thomas Colby, May 16, 1842. Papers on the tides, 1839–1844, RGO6/499: 461–62. Papers of George Biddell Airy, University of Cambridge Library.

———. "On the Laws of Individual Tides at Southampton and at Ipswich." *Philosophical Transactions of the Royal Society of London* 133 (1843): 45–54.

———. "On the Laws of the Tides on the Coasts of Ireland, as Inferred from an Extensive Series of Observations Made in Connection with the Ordnance Survey of Ireland." *Philosophical Transactions of the Royal Society of London* 135 (1845): 1–124.

Aiton, E. J. "Descartes's Theory of the Tides." *Annals of Science* 11, no. 4 (1955): 337–48.

Albrecht, Glenn. "'Solastalgia': A New Concept in Health and Identity." *Philosophy Activism Nature*, no. 3 (2005): 41–55.

Albrecht, Glenn, Gina-Maree Sartore, Linda Connor, Nick Higginbotham, Sonia Freeman, Brian Kelly, Helen Stain, Anne Tonna, and Georgia Pollard. "Solastalgia: The Distress Caused by Environmental Change." *Australasian Psychiatry* 15, no. 1, suppl. (2007): S95–98.

Alexander, Ralph B. "No Evidence That Climate Change Is Accelerating Sea Level Rise." *Science under Attack* (blog), September 23, 2019. https://www.scienceunderattack.com/blog/2019/9/23/no-evidence-that-climate-change-is-accelerating-sea-level-rise-35.

"Andaman, Nicobar Islands May Not Be Inhabitable in Future Due to Rise in Sea Level: IPCC." *Times of India*, September 25, 2019. https://timesofindia.indiatimes.com/india/andaman-nicobar-islands-may-not-be-inhabitable-in-future-due-to-rise-in-sea-level-ipcc/articleshow/71298163.cms.

Annual Report of the Superintendent, United States Coast and Geodetic Survey to the Secretary of Commerce for the Fiscal Year Ended June 30, 1917. Washington, DC: Government Printing Office, 1917.

Arago, François. "Notices scientifiques." *Annuaire du Bureau des longitudes*, 1836, 189–349.

———. "Sur les phénomènes de la mer." In *Oeuvres complètes*, edited by Jean Augustin Barral, 9:549–635. Paris: Gide, 1857.

Argonaut. "Permanent Difference of Level in Different Parts of the Ocean Considered." *Nautical Magazine and Naval Chronicle* 6 (1837): 306–13; 384–91.

Ariza, Mario Alejandro. "As Miami Keeps Building, Rising Seas Deepen Its Social Divide." *YaleEnvironment360*, September 29, 2020. https://e360.yale.edu/features/as-miami-keeps-building-rising-seas-deepen-its-social-divide.

Arnu, Titus. "8848,86." *Süddeutsche Zeitung*, December 8, 2020. https://www.sueddeutsche.de/panorama/mount-everest-vermessung-neue-hoehe-86-zentimeter-mehr-1.5141203.

Association d'Océanographie Physique. *General Assembly at Edinburgh. September 1936*. Procès-Verbaux 2. Liverpool: Secrétariat de l'Association, 1937.

———. *Monthly and Annual Mean Heights of Sea-Level, Up to and Including the Year 1936*. Publication Scientifique 5. Liverpool: Secrétariat de l'Association, 1939.

———. *Monthly and Annual Mean Heights of Sea-Level 1937 to 1946 and Unpublished Data for Earlier Years*. Publication Scientifique 10. Oslo: Secrétariat de l' Association, 1950.

———. *Monthly and Annual Mean Heights of Sea-Level 1947 to 1951 and Unpublished Data for Earlier Years*. Publication Scientifique 12. Oslo: Secrétariat de l' Association, 1953.

———. *Secular Variation of Sea-Level*. Publication Scientifique 13. Bergen: Secrétariat de l'Association, 1954.

Baeyer, Johann Jakob. *Nivellement zwischen Swinemünde und Berlin*. Berlin: Ferdinand Dümmler, 1840.

———. *Zur Entstehungsgeschichte der europäischen Gradmessung*. Berlin: Stankiewicz, 1862.

Baker, Victor R. "Catastrophism and Uniformitarianism: Logical Roots and Current Relevance in Geology." In *Lyell: The Past Is the Key to the Present*, edited by D. J. Blundell and A. C. Scott, 171–82. Special Publications 143. London: Geological Society, 1998.

Ballard, J. G. *The Drowned World*. New York: Berkley, 1962.

Barnett, Lydia. *After the Flood: Imagining the Global Environment in Early Modern Europe*. Baltimore: Johns Hopkins University Press, 2019.

———. "The Theology of Climate Change: Sin as Agency in the Enlightenment's Anthropocene." *Environmental History* 20, no. 2 (2015): 217–37.

Bartky, Ian R. *One Time Fits All: The Campaigns for Global Uniformity*. Stanford, CA: Stanford University Press, 2007.

Bateman, J. F. "Some Account of the Suez Canal: In a Letter to the President." *Proceedings of the Royal Society*, no. 116 (1870).

Baulig, Henri. *The Changing Sea Level: Four Lectures Given at the University of London in November 1933*. Institute of British Geographers Special Publication 3. 1935. Reprint, London: George Philip & Son, 1956.

Beautemps-Beaupré, Charles François, and Pierre Daussy. *Exposé des travaux relatifs à la reconnaissance hydrographique des côtes occidentales de France*. Paris: Imprimerie Royale, 1829.

Beck, David. "Rechenfehler beim Bau der Hochrheinbrücke." *SWR Wissen*, January 9, 2023. https://www.swr.de/wissen/flops-in-der-technikgeschichte-hochrheinbruecke-100.html.

Benson, Etienne. *Surroundings: A History of Environments and Environmentalisms*. Chicago: University of Chicago Press, 2020.

Berghaus, Heinrich. "Alexander von Humboldt's System der Isotherm-Kurven in Merkator's Projection." In *Dr. Heinrich Berghaus' Physikalischer Atlas*, 2nd ed. Vol. 1. Gotha: J. Perthes, 1849.

Bessel, Friedrich Wilhelm. "Ueber den Einfluss der Unregelmässigkeiten der Figur der Erde, auf geodätische Arbeiten und ihre Vergleichung mit den astronomischen Bestimmungen." *Astronomische Nachrichten* 14, nos. 19–21 (1837): 269–312.

Bevan, Benjamin. Letter to Admiralty Office, June 21, 1823. Miscellaneous Tides and Trade Winds, RGO 14/51: 112–13. Board of Longitude Papers, University of Cambridge Library. https://cudl.lib.cam.ac.uk/view/MS-RGO-0001400051/242.

———. Letter to William Hyde Wollaston, July 22, 1822. Miscellaneous Tides and Trade Winds, RGO 14/51: 88–90. Board of Longitude Papers, University of Cambridge Library. https://cudl.lib.cam.ac.uk/view/MS-RGO-0001400051/183.

Bhattacharyya, Debjani. *Empire and Ecology in the Bengal Delta: The Making of Calcutta*. Studies in Environment and History. Cambridge: Cambridge University Press, 2018.

———. "A River Is Not a Pendulum: Sediments of Science in the World of Tides." *Isis* 112, no. 1 (2021): 141–49.

Biasillo, Roberta. "Historical Tools and Current Societal Challenges: Reflections on a Collection of Environmental Migration Cases." *Fennia—International Journal of Geography* 198, nos. 1–2 (2020): 151–62.

Bińczyk, Ewa. "The Most Unique Discussion of the 21st Century? The Debate on the Anthropocene Pictured in Seven Points." *Anthropocene Review* 6, nos. 1–2 (2019): 3–18.

Biswas, Sravani, and Patrick Daly. "'Cyclone Not above Politics': East Pakistan, Disaster Politics, and the 1970 Bhola Cyclone." *Modern Asian Studies* 55, no. 4 (2021): 1382–1410.

Blanchard, Raoul. "M. Auguste Bouchayer." *Revue de géographie alpine* 31, no. 2 (1943): 249–50.

Bloch, Marc. *The Historian's Craft*. 1949. Reprint, New York: Vintage Books, 1953.

Bloch, Moshe Rudolph. "A Hypothesis for the Change of Ocean Levels Depending on the Albedo of the Polar Ice Caps." *Palaeogeography, Palaeoclimatology, Palaeoecology* 1 (1965): 127–42.

Börsch, Anton. "Vergleichung der Mittelwasser und der Nullpunkte für die Höhen." In *Verhandlungen der vom 27. September bis 7. October 1892 in Brüssel abgehaltenen zehnten allgemeinen Conferenz der Internationalen Erdmessung*, edited by Adolph Hirsch, Beilage A VI:547–52. Berlin: Reimer, 1893.

Bouchayer, Auguste. *Marseille; ou, la mer qui monte*. Paris: Les Editions des Portiques, 1931.

Bourdalouë, Paul-Adrien. "Notice sur le nivellement de l'isthme de Suez et de la Basse Égypte," March 7, 1847. Fonds Enfantin: Ms-7832. Bibliothèque de l'Arsenal (BNF). https://gallica.bnf.fr/ark:/12148/btv1b52509327j.

Bourdiol, Henri. "Importance d'un nivellement général de la France et opportunité d'en assurer l'execution." *Bulletin de la Société de géographie*, 5, no. 10 (1865): 177–96.

Bowie, William. "Notable Progress in Surveying Instruments." *Scientific Monthly* 29, no. 5 (1929): 402–6.

Bradshaw, E., P. L. Woodworth, A. Hibbert, L. J. Bradley, D. T. Pugh, C. Fane, and R. M. Bingley. "A Century of Sea Level Measurements at Newlyn, Southwest England." *Marine Geodesy* 39, no. 2 (2016): 115–40.

Broc, Numa. *Les montagnes au Siècle des lumières: Perception et représentation*. Ebook. Mémoires de la Section de géographie physique et humaine 4. Paris: Editions du CTHS, 1991.

Broecker, Wallace S. "Climatic Change: Are We on the Brink of a Pronounced Global Warming?" *Science* 189, no. 4201 (1975): 460–63.

Bruchhausen, Wilhelm von. *Die periodisch wiederkehrenden Eiszeiten und Sindfluten und die wichtigsten Folgerungen aus diesen wechselnden Ueberschwemmungen der südlichen und der nördlichen Kontinente*. Trier: C. Troschel, 1845.

Bruhns, Carl, Wilhelm Förster, and Adolph Hirsch, eds. *Bericht über die Verhandlungen der vom 30. September bis 7. October 1867 zu Berlin abgehaltenen allgemeinen Conferenz der Europäischen Gradmessung*. Berlin: Reimer, 1868.

Bruhns, Carl, and Adolph Hirsch, eds. *Bericht über die Verhandlungen der vom 21. bis 30. September 1871 zu Wien abgehaltenen dritten allgemeinen Conferenz der Europäischen Gradmessung*. Berlin: Reimer, 1872.

———. *Verhandlungen der vom 20. bis 29. September 1875 in Paris vereinigten Permanenten Commission der Europäischen Gradmessung*. Berlin: Reimer, 1876.

———. *Verhandlungen der vom 5. bis 10. October 1876 in Brüssel vereinigten Permanenten Commission der Europaeischen Gradmessung*. Berlin: Reimer, 1877.

Bruncrona, Nils, and Carl P. Hällström. "Beobachtungen und Angaben über die Verminderung des Wassers an der Schwedischen Küste." *Annalen der Physik und Chemie* 2, no. 11 (1824): 308–28.

Buch, Leopold von. *Travels through Norway and Lapland during the Years 1806, 1807, and 1808*. Translated by John Black. London: Henry Colburn, 1813.

Buhrman, Kristina. "Remembering Future Risk: Considering Technologies of the Archive for Discussion of Tōhoku's Seismological Past after 2011." *Technology and Culture* 58, no. 1 (2017): 159–69.

Burstyn, Harold L. "Seafaring and the Emergence of American Science." In *The Atlantic World of Robert G. Albion*, edited by Benjamin W. Labaree, 76–109. Middletown, CT: Wesleyan University Press, 1975.

Butzengeiger, Sonja, and Britta Horstmann. *Sea-Level Rise in Bangladesh and the Netherlands: One Phenomenon, Many Consequences*. Bonn: Germanwatch, 2004.

Cajori, Florian. "History of Determinations of the Heights of Mountains." *Isis* 12, no. 3 (1929): 482–514.

Cariou, Gautier. "Où se trouve le niveau de la mer?" *La Recherche*, November 2016. https://www.larecherche.fr/où-se-trouve-le-niveau-de-la-mer.

Carson, Rachel. *The Edge of the Sea*. Boston: Houghton Mifflin, 1955.

Carter, Paul. *Dark Writing: Geography, Performance, Design*. Honolulu: University of Hawai'i Press, 2008.

Cartwright, David E. "The Historical Development of Tidal Science, and the Liverpool Tidal Institute." In *Oceanography: The Past*, edited by Mary Sears and Daniel Merriman, 240–51. New York: Springer, 1980.

———. *Tides: A Scientific History*. Cambridge: Cambridge University Press, 1999.

Cazenave, A., N. Champollion, J. Benveniste, and P. Lecomte. "International Space Science Institute (ISSI) Workshop on Integrative Study of the Mean Sea Level and Its Components." In *Integrative Study of the Mean Sea Level and Its Components*, edited by Anny Cazenave, Nicolas Champollion, Frank Paul, and Jérôme Benveniste, 1–5. Space Sciences Series of ISSI. Cham: Springer International Publishing, 2017.

Celsius, Andreas. "Anmärkning om vatnets förminskande så i Östersjön som Vesterhafvet." *Kongl. Swenska Wetenskaps Academiens Handlingar* 4 (1743): 33–50.

"Centenary of the Nautical Magazine." *Nature* 129, no. 3248 (1932): 163–63.

Chakrabarti, Pratik. *Inscriptions of Nature: Geology and the Naturalization of Antiquity*. Baltimore: Johns Hopkins University Press, 2020.

Chakrabarty, Dipesh. "The Climate of History: Four Theses." *Critical Inquiry* 35, no. 2 (2009): 197–222.

———. *The Climate of History in a Planetary Age*. Chicago: University of Chicago Press, 2021.

Chambers, Charles. *The Meteorology of the Bombay Presidency*. London: Her Majesty's Stationery Office, 1878.

Chambers, Robert. *Ancient Sea-Margins, as Memorials of Changes in the Relative Level of Sea and Land*. Edinburgh: W. & R. Chambers, 1848.

Chaumont, Jean-Philippe. "Ministére des Travaux publics; Nivellement général de la France dit Bourdaloué; Carnets des opérateurs et des lecteurs (1855–1863); F/14/4583 a 5672." Archives Nationales (Paris), 2009.

Chevalier, Michel. "L'isthme de Panama." *Revue des Deux Mondes*, year 14, n.s., 5 (1844): 5–74.

Church, J. A., J. M. Gregory, P. Huybrechts, M. Kuhn, K. Lambeck, M. T. Nhuan, D. Qin, and P. L Woodworth. "Changes in Sea Level." In *Climate Change 2001: The Scientific Basis*, edited by John Theodore Houghton, Yihui Ding, D. J. Griggs, M. Noguer, P. J. van der Linden, X. Dai, K. Maskell, and C. A. Johnson, 639–93. New York: Cambridge University Press, 2001.

Cliffe, Charles Frederick. *The Book of South Wales, the Bristol Channel, Monmouthshire and the Wye*. 2nd ed. London: Hamilton, Adams & Company, 1848.

Climate Central. "Comparison: Long-Term Sea Level Outcomes." Accessed October 14, 2021. https://coastal.climatecentral.org/map/6/5.6552/51.5046/?theme=warming&map_type=multicentury_slr_comparison&basemap=roadmap&elevation_model=best_available&lockin_model=levermann_2013&temperature_unit=C&warming_comparison=["1.5"%2C"3.0"].

"Climate Changes and the Last Glacial Period." *Science* 77, no. 1993 (1933): 9.

Close, Charles, ed. *The Second Geodetic Levelling of England and Wales 1912–1921*. London: His Majesty's Stationery Office, 1922.

Cock, A. G. "Chauvinism and Internationalism in Science: The International Research Council, 1919–1926." *Notes and Records of the Royal Society of London* 37, no. 2 (1983): 249–88.

Cohen, Darryl. "About 60.2M Live in Areas Most Vulnerable to Hurricanes." United States Census Bureau, July 15, 2019. https://www.census.gov/library/stories/2019/07/millions-of-americans-live-coastline-regions.html.

Conrad, Sebastian. *Deutsche Kolonialgeschichte*. Munich: C. H. Beck, 2008.

Cook, Alan Hugh. "Determination of the Earth's Gravitational Potential from Observations of Sputnik 2 (1957b)." *Geophysical Journal of the Royal Astronomical Society* 1, no. 4 (1958): 341–45.

Coraboeuf, Jean-Baptiste. "Appendice. Exposé des opérations qui ont été faites, en 1825, aux deux extrémités de la base de Perpignan." In *Mémoires présentés par divers savans a l'Académie Royale des Sciences de l'Institut de France*, 3:105–31. Paris: Académie Royale des Sciences, 1832.

———. "Mémoire sur les opérations géodésiques des Pyrénées et la comparaison du niveau des deux mers." In *Mémoires présentés par divers savans a l'Académie Royale des Sciences de l'Institut de France*, 3:45–104. Paris: Académie Royale des Sciences, 1832.

Corbin, Alain. *The Lure of the Sea: The Discovery of the Seaside in the Western World, 1750–1840*. Translated by Jocelyn Phelps. 1988. Reprint, Berkeley: University of California Press, 1994.

Correa de Serra, Joseph. "On a Submarine Forest, on the East Coast of England." *Philosophical Transactions of the Royal Society of London* 89 (1799): 145–56.

Coulomb, Alain. *Le marégraphe de Marseille: de la détermination de l'origine des altitudes au suivi des changements climatiques 130 ans d'observation du niveau de la mer*. Paris: Presses des Ponts, 2014.

Croll, James. *Climate and Time in Their Geological Relations: A Theory of Secular Changes of the Earth's Climate*. New York: D. Appleton and Company, 1875.

———. "On the Change in the Obliquity of the Ecliptic: Its Influence on the Climate of the Polar Regions, and the Level of the Sea." *Transactions of the Geological Society of Glasgow* 2 (1867): 177–98.

———. "On the Physical Cause of the Submergence of the Land during the Glacial Epoch." Edited by David Masson. *The Reader* 6, no. 140 (1865): 270–71.

Crouzet-Pavan, Elizabeth. "Venice and Its Surroundings." In *A Companion to Venetian History, 1400–1797*, edited by Eric R. Dursteler, 25–46. Leiden: Brill, 2013.

Cusick, Daniel. "Miami Is the 'Most Vulnerable' Coastal City Worldwide." *Scientific American*, February 4, 2020. https://www.scientificamerican.com/article/miami-is-the-most-vulnerable-coastal-city-worldwide/.

Daly, Reginald A. "The Glacial-Control Theory of Coral Reefs." *Proceedings of the American Academy of Arts and Sciences* 51, no. 4 (1915): 157–251.

———. "A Recent Worldwide Sinking of Ocean-Level." *Geological Magazine* 57, no. 6 (1920): 246–61.

Dam, Petra J. E. M. van. *Van Amsterdams Peil naar Europees referentievlak: de geschiedenis van het NAP tot 2018*. Hilversum: Verloren, 2018.

Darwin, Charles. *Geological Observations on Coral Reefs, Volcanic Islands, and on South America: Being the Geology of the Voyage of the Beagle, under the Command of Captain Fitzroy, R.N., during the Years 1832 to 1836*. 1842. Reprint, London: Smith, Elder & Co., 1851.

———. *The Structure and Distribution of Coral Reefs: Being the First Part of the Geology of the Voyage of the Beagle, under the Command of Capt. Fitzroy, R.N. during the Years 1832 to 1836*. London: Smith, Elder & Co., 1842.

Darwin, George H. "The British Empire." In *Verhandlungen der vom 3. bis 12. October 1898 in Stuttgart abgehaltenen zwölften Allgemeinen Conferenz der Internationalen Erdmessung*, edited by Adolph Hirsch, Annexes B XXI. Berlin: Georg Reimer, 1899.

"Das Niveau der Ostsee. Nach den Messungen der Königl. Preussischen Landes-Aufnahme." *Mittheilungen aus Justus Perthes' Geographischer Anstalt über wichtige neue Erforschungen auf dem Gesammtgebiete der Geographie von Dr. A. Petermann* 21 (1875): 229–30.

Daussy, Pierre. "Mémoire sur les marées des côtes de France." *Annales maritimes et coloniales* 17 (1832): 558–75.

Davidson, George. "États-Unis." In *Verhandlungen der vom 3. bis 12. October 1889 in Paris abgehaltenen neunten allgemeinen Conferenz der Internationalen Erdmessung und deren Permanenten Commission*, edited by Adolph Hirsch. Annexe B XVII. Berlin: Georg Reimer, 1890.

Davies, Bethan. "Calculating Glacier Ice Volumes and Sea Level Equivalents." *AntarcticGlaciers.org* (blog), July 15, 2020. http://www.antarcticglaciers.org/glaciers-and-climate/estimating-glacier-contribution-to-sea-level-rise/.

———. "If All the Ice in Antarctica Were to Melt, How Much Would Global Sea Level Rise? How Quickly Is This Likely to Happen?" *AntarcticGlaciers.org* (blog), October 1, 2013. http://www.antarcticglaciers.org/question/ice-antarctica-melt-much-global-sea-level-rise-quickly-likely-happen/.

Davies, Gordon L. *The Earth in Decay: A History of British Geomorphology*. London: Macdonald & Co., 1969.

Dawson, W. Bell. *Tide Levels and Datum Planes in Eastern Canada: From Determinations by the Tidal and Current Survey up to the Year 1917*. Ottawa: Department of the Naval Service, 1917.

De Geer, Gerard. "Om Skandinaviens nivåförändringar under qvartärperioden." *Geologiska Föreningen i Stockholm Förhandlingar* 10, no. 5 (1888): 366–79.

Deacon, Margaret. *Scientists and the Sea, 1650–1900: A Study of Marine Science*. London: Academic Press, 1971.

Denham, Henry Mangles. "On the Survey of the Mersey and the Dee." In *Notices and Communications to the British Association for the Advancement of Science at Dubln in August 1835*, 64–66. London: John Murray, 1836.

Deparis, Vincent, Hilaire Legros, and Jean Souchay. "Investigations of Tides from the Antiquity to Laplace." In *Tides in Astronomy and Astrophysics*, edited by Jean Souchay, Stéphane Mathis, and Tadashi Tokieda, 861:31–82. Berlin: Springer, 2013.

"Der Normal-Höhenpunkt für das Königreich Preußen." *Zeitschrift für Vermessungswesen* 9, no. 1 (1880): 1–16.

Der Normal-Höhepunkt für das Königreich Preussen an der königlichen Sternwarte zu Berlin. Berlin: Trigonometrische Abtheilung der Landesaufnahme, 1879.

Desmarest, Nicolas. "Ferner. Précis de la discussion qui a eu lieu entre les savans de Suède & d'Italie, sur la diminution des eaux de la mer & ses progrès." In *Encyclopédie méthodique. Géographie-physique*, 1:133–50. Paris: H. Agasse, 1795.

Diderot, Denis, ed. "Montagnes." In *Encyclopédie; ou, Dictionnaire raisonné des Sciences, des arts et des métiers*, 10:676–79. Neuchatel: Samuel Faulche et Compagnie, 1765.

"Disappearing World: Global Warming Claims Tropical Island." *Independent*, December 24, 2006, https://web.archive.org/web/20061228004107/http://news.independent.co.uk/environment/article2099971.ece.

"Discours prononcé par M. Alexandre de Humboldt à la sèance extraordinaire de l'Academie impériale des sciences de Saint-Pétersbourg, tenue le 16/28 novembre 1829." *Nouvelles Annales des Voyages* 2, no. 15 (1830): 86–101.

Dongo, Alexandre. "Le volcan Chimborazo plus haut que l'Everest . . ." *Equateur Info* (blog), April 12, 2016. https://www.equateur.info/volcan-chimborazo-haut-leverest/.

Donn, William L., and David M. Shaw. "Sea Level and Climate of the Past Century." *Science* 142, no. 3596 (1963): 1166–67.

Dossal, Mariam. *Theatre of Conflict, City of Hope: Mumbai, 1660 to Present Times*. New Delhi: Oxford University Press, 2010.

Dott, Robert H., ed. *Eustasy: The Historical Ups and Downs of a Major Geological Concept*. Memoir 180. Boulder, CO: Geological Society of America, 1992.

Droxler, André W., and Stéphan J. Jorry. "The Origin of Modern Atolls: Challenging Darwin's Deeply Ingrained Theory." *Annual Review of Marine Science* 13, no. 1 (2021): 21.1–37.

Dumont, J. P., V. Rosmorduc, N. Picot, E. Bronner, S. Desai, H. Bonekamp, J. Figa, J. Lillibridge, and R. Scharoo. *OSTM/Jason-2 Products Handbook*. Version 1, rev. 8. Paris: CNES, 2011. https://www.ospo.noaa.gov/Products/documents/J2_handbook_v1-8_no_rev.pdf.

Dunn, Richard, and Rebekah Higgitt. *Finding Longitude: How Ships, Clocks and Stars Helped Solve the Longitude Problem*. London: Collins and Royal Museums Greenwich, 2014.

Dutton, Clarence Edward. "On Some of the Greater Problems of Physical Geology." *Bulletin of the Philosophical Society of Washington* 11 (1892): 51–64.

Edwards, Paul N. *A Vast Machine: Computer Models, Climate Data, and the Politics of Global Warming*. Cambridge, MA: MIT Press, 2010.

Ekman, Martin. "A Concise History of Postglacial Land Uplift Research (from Its Beginning to 1950)." *Terra Nova* 3, no. 4 (1991): 358–65.

Elliott, James R., and Jeremy Pais. "Race, Class, and Hurricane Katrina: Social Differences in Human Responses to Disaster." *Social Science Research* 35, no. 2, "Katrina in New Orleans/Special Issue on Contemporary Research on the Family" (2006): 295–321.

Englander, John. "Does Flood Insurance Cover Rising Seas?" *Sea Level Rise Now* (blog), March 17, 2021. https://johnenglander.net/does-flood-insurance-cover-rising-seas/.

EPA. *Climate Change Indicators in the United States*. 4th ed. EPA 430-R16-004. Washington, DC: US Environmental Protection Agency, 2016. https://www.epa.gov/sites/ production/files/2016-08/documents/climate_ indicators_2016.pdf.

Etkins, Robert, and Edward S. Epstein. "The Rise of Global Mean Sea Level as an Indication of Climate Change." *Science* 215, no. 4530 (1982): 287–89.

EUMETSAT. "Jason-3 Instruments," May 26, 2020. https://www.eumetsat.int/jason-3 -instruments.

Ewing, Gifford C., ed. *Oceanography from Space: Proceedings of Conference on the Feasability of Conducting Oceanographic Explorations from Aircraft, Manned Orbital and Lunar Laboratories: Held at Woods Hole, Massachusetts, 24–28 August 1964*. Woods Hole Oceanographic Institution, 1965.

Eyles, V. A. "Hutton, James (*b*. Edinburgh, Scotland, 3 June 1726; *d*. Edinburgh, 26 March 1797)." In *Dictionary of Scientific Biography*, edited by Charles S. Gillespie, 120–32. New York: Linda Hall Library, 1971. https://www.chlt.org/sandbox/lhl/dsb/page.4.a.php.

Faber, Jacob William. "Superstorm Sandy and the Demographics of Flood Risk in New York City." *Human Ecology* 43, no. 3 (2015): 363–78.

Fairbridge, Rhodes W. "The Changing Level of the Sea." *Scientific American* 202, no. 5 (1960): 70–79.

———. "Dating the Latest Movements of the Quaternary Sea Level." *Transactions of the New York Academy of Sciences* 20, no. 6, series II (1958): 471–82.

———. "Eustatic Changes in Sea Level." *Physics and Chemistry of the Earth* 4 (1961): 99–185.

———. "Raised Beach." In *Beaches and Coastal Geology*, edited by M. Schwartz, 677–78. Encyclopedia of Earth Sciences Series. New York, NY: Springer US, 1982.

Feldman, Theodore S. "Applied Mathematics and the Quantification of Experimental Physics: The Example of Barometric Hypsometry." *Historical Studies in the Physical Sciences* 15, no. 2 (1985): 127–95.

Fisher, James F. *Sherpas: Reflections on Change in Himalayan Nepal*. Berkeley: University of California Press, 1990.

Flannery, Tim. *The Weather Makers: The History and Future Impact of Climate Change*. Melbourne: Text Publishing, 2005.

Fleetwood, Lachlan. "Bodies in High Places: Exploration, Altitude Sickness, and the Problem of Bodily Comparison in the Himalaya, 1800–1850." *Itinerario* 43, no. 3 (2019): 489–515.

———. "'No Former Travellers Having Attained Such a Height on the Earth's Surface': Instruments, Inscriptions, and Bodies in the Himalaya, 1800–1830." *History of Science* 56, no. 1 (2018): 3–34.

Fleming, James Rodger. "Charles Lyell and Climatic Change: Speculation and Certainty." In *Lyell: The Past Is the Key to the Present*, edited by D. J. Blundell and A. C. Scott, 161–69. Special Publications 143. London: Geological Society, 1998.

———. "The Pathological History of Weather and Climate Modification: Three Cycles of Promise and Hype." *Historical Studies in the Physical and Biological Sciences* 37, no. 1 (2006): 3–25.

"Flood Map: Elevation Map, Sea Level Rise Map." Accessed November 9, 2021. https://www.floodmap.net/.

Förster, Wilhelm, ed. *Verhandlungen der ersten allgemeinen Conferenz der Bevollmächtigten zur Mittel-Europäischen Gradmessung vom 15. bis 22. October 1864*. Berlin: Königlichen Geheimen Ober-Hofbuchdruckerei, 1865.

Franta, Benjamin. "On Its 100th Birthday in 1959, Edward Teller Warned the Oil Industry about Global Warming." *Guardian*, January 1, 2018. http://www.theguardian.com/environment/climate-consensus-97-per-cent/2018/jan/01/on-its-hundredth-birthday-in-1959-edward-teller-warned-the-oil-industry-about-global-warming.

Gartner, William. "Mapmaking in the Central Andes." In *The History of Cartography*, vol. 2, book 3, edited by David Woodward and G. Malcolm Lewis, 257–300. Chicago: University of Chicago Press, 1998.

Geiger, Reed. "Planning the French Canals: The 'Becquey Plan' of 1820–1822." *Journal of Economic History* 44, no. 2 (1984): 329–39.

Geyer, Martin H. "One Language for the World: The Metric System, International Coinage, Gold Standard, and the Rise of Internationalism, 1850–1900." In *The Mechanics of Internationalism: Culture, Society, and Politics from the 1840s to the First World War*, edited by Martin H. Geyer and Johannes Paulmann, 55–92. New York: Oxford University Press, 2001.

Ghosh, Amitav. *The Great Derangement: Climate Change and the Unthinkable*. Chicago: University of Chicago Press, 2017.

———. *The Nutmeg's Curse: Parables for a Planet in Crisis*. London: John Murray, 2021.

Goby, Jean-Edouard. "Histoire des nivellements de l'Isthme de Suez." *Bulletin de la Sociètè d'Études Historiques et Géographiques de l'Isthme de Suez* 4 (1952): 99–170.

———. "Marées de la Mer Rouge à Port-Taufiq et de la Méditerranée à Port-Saïd." *Bulletin de la Sociètè d'Études Historiques et Géographiques de l'Isthme de Suez* 3 (1951): 17–32.

Godwin, H. "Coastal Peat Beds of the British Isles and North Sea: Presidential Address to the British Ecological Society 1943." *Journal of Ecology* 31, no. 2 (1943): 199–247.

Godwin, H., R. P. Suggate, and E. H. Willis. "Radiocarbon Dating of the Eustatic Rise in Ocean-Level." *Nature* 181, no. 4622 (May 1958): 1518–19.

Goüye, P., and M. de la Hire. "Memoire de la maniere d'observer dans les ports le flux et le reflux de la mer." In *Histoire de l'Académie royale des sciences. Année MDCCI*, 12–13. Paris: Gabriel Martin, Jean-Bat. Coignard, & Hyppolite-Louis Guerin, 1743.

Greenaway, Frank. *Science International: A History of the International Council of Scientific Unions*. Cambridge: Cambridge University Press, 1996.

Greenough, George B. "Address Delivered at the Anniversary Meeting, 21 February 1834." *Proceedings of the Geological Society of London* 2, no. 35 (1838): 42–70.

Grosjean, Georges. *Geschichte der Kartographie*. 3rd ed. Geographica Bernensia U8. Bern: Geographisches Institut der Universität Bern, 1996.

Gudrais, Elizabeth. "The Gravity of Glacial Melt." *Harvard Magazine*, May 1, 2010. https://www.harvardmagazine.com/2010/05/gravity-of-glacial-melt.

Guldi, Jo. *Roads to Power: Britain Invents the Infrastructure State*. Cambridge, MA: Harvard University Press, 2012.

Gutenberg, Beno. "Changes in Sea Level, Postglacial Uplift, and Mobility of the Earth's Interior." *Geological Society of America Bulletin* 52, no. 5 (May 1, 1941): 721–72.

Hafeneder, Rudolf. "Deutsche Kolonialkartographie 1884–1919." PhD diss., Universität der Bundeswehr, 2008. http://athene-forschung.unibw.de/doc/86117/86117.pdf.

Haith, Chelsea. "Sea-Level Rise: Writers Imagined Drowned Worlds for Centuries—What They Tell Us about the Future." *The Conversation*, January 28, 2021. http://theconversation.com/sea-level-rise-writers-imagined-drowned-worlds-for-centuries-what-they-tell-us-about-the-future-151804.

Hamblin, Jacob Darwin. *Oceanographers and the Cold War: Disciples of Marine Science*. Seattle: University of Washington Press, 2005.

Hann, Julius von. "Ueber gewisse beträchtliche Unregelmässigkeiten des Meeres-Niveaus." *Mittheilungen der kaiserlich-königlichen Geographischen Gesellschaft* 18 (1875): 554–69.

Hardenberg, Wilko Graf von. "Making a Stable Sea: The Littorals of Eighteenth-Century Europe and the Origins of a Spatial Concept." *Isis* 112, no. 1 (March 1, 2021): 130–40.

———. "Measuring Zero at Sea: On the Delocalization and Abstraction of the Geodetic Framework." *Journal of Historical Geography* 68 (April 1, 2020): 11–20.

Hardenberg, Wilko Graf von, and Martin Mahony. "Introduction—Up, Down, Round and Round: Verticalities in the History of Science." *Centaurus* 62, no. 4 (2020): 595–611.

Hauer, Mathew E., Elizabeth Fussell, Valerie Mueller, Maxine Burkett, Maia Call, Kali Abel, Robert McLeman, and David Wrathall. "Sea-Level Rise and Human Migration." *Nature Reviews Earth and Environment* 1, no. 1 (January 2020): 28–39.

Helmert, Friedrich Robert. "Le zéro des altitudes." In *Verhandlungen der Conferenz der Permanenten Commission der Internationalen Erdmessung/8. bis 17. October 1891 zu Florenz*, edited by Adolph Hirsch, 148–53. Berlin: Reimer, 1892.

Henrici, E. O. "Mean Sea-Level." *Geographical Journal* 38, no. 6 (1911): 605–6.

Herren, Madeleine. "Governmental Internationalism and the Beginning of a New World Order in the Late Nineteenth Century." In *The Mechanics of Internationalism: Culture, Society, and Politics from the 1840s to the First World War*, edited by Martin H. Geyer and Johannes Paulmann, 121–44. New York: Oxford University Press, 2001.

Hestmark, Geir. "Tracings of the North of Europe: Robert Chambers in Search of the Ice Age." *Annals of Science* 74, no. 4 (2017): 262–81.

Heyde, Herbert. *Die Höhennullpunkte der amtlichen Kartenwerke der europäischen Staaten und ihre Lage zu Normal-Null*. 1923. Reprint, Dortmund: Förderkreis Vermessungstechn. Museum, 1999.

Hirsch, Adolph. "Rapport sur l'état actuel des travaux de nivellement de précision exécutés dans les différents pays de l'Association." In *Verhandlungen der vom 15. bis zum 24. October 1883 in Rom abgehaltenen siebenten allgemeinen Conferenz der Europäischen Gradmessung*, edited by Adolph Hirsch, Annexe IV. Berlin: Georg Reimer, 1884.

———, ed. *Verhandlungen der vom 27. October bis zum 1. November 1886 in Berlin abgehaltenen achten allgemeinen Conferenz der Internationalen Erdmessung und deren Permanenten Commission*. Berlin: Georg Reimer, 1887.

———, ed. *Verhandlungen der vom 21. bis zum 29. October 1887 auf der Sternwarte zu Nizza abgehaltenen Conferenz der Permanenten Commission der Internationalen Erdmessung*. Berlin: Reimer, 1888.

———, ed. *Verhandlungen der vom 17–23. September 1888 in Salzburg abgehaltenen Conferenz der Permanenten Commission der Internationalen Erdmessung*. Berlin: Reimer, 1889.

———, ed. *Verhandlungen der vom 3. bis 12. October 1889 in Paris abgehaltenen neunten allgemeinen Conferenz der Internationalen Erdmessung und deren Permanenten Commission*. Berlin: Georg Reimer, 1890.

———, ed. *Verhandlungen der vom 15. bis 21. September 1890 zu Freiburg i/B. abgehaltenen Conferenz der Permanenten Commission der Internationalen Erdmessung*. Berlin: Reimer, 1891.

———, ed. *Verhandlungen der vom 8. bis 17. October 1891 zu Florenz abgehaltenen Conferenz der Permanenten Commission der Internationalen Erdmessung*. Berlin: Reimer, 1892.

———, ed. *Verhandlungen der vom 27. September bis 7. October 1892 in Brüssel abgehaltenen zehnten allgemeinen Conferenz der Internationalen Erdmessung und deren Permanenten Commission*. Berlin: Reimer, 1893.

Hirsch, Adolph, and Theodor von Oppolzer, eds. *Verhandlungen der vom 11. bis zum 15. September 1882 im Haag vereinigten Permanenten Commission der Europäischen Gradmessung*. Berlin: Reimer, 1883.

Hodgson, J. A., and J. D. Herbert. "An Account of Trigonometrical and Astronomical Operations for Determining the Heights and Positions of the Principal Peaks of the Himalaya Mountains." *Asiatick Researches* 14 (1822): 187–372.

Hoffman, John S., Dale Keyes, and James G. Titus. "Projecting Future Sea Level Rise: Methodology, Estimates to the Year 2100, & Research Needs." United States Environmental Protection Agency, October 1983.

Höhler, Sabine, and Nina Wormbs. "Remote Sensing: Digital Data at a Distance." In *Methodological Challenges in Nature-Culture and Environmental History Research*, edited by Jocelyn Thorpe, Stephanie Rutherford, and L. Anders Sandberg. London: Routledge, 2016.

Holgate, Simon J., Andrew Matthews, Philip L. Woodworth, Lesley J. Rickards, Mark E. Tamisiea, Elizabeth Bradshaw, Peter R. Foden, Kathleen M. Gordon, Svetlana Jevrejeva, and Jeff Pugh. "New Data Systems and Products at the Permanent Service for Mean Sea Level." *Journal of Coastal Research* 29, no. 3 (2013): 493–504.

Houghton, Katherine J., Athanasios T. Vafeidis, Barbara Neumann, and Alexander Proelss. "Maritime Boundaries in a Rising Sea." *Nature Geoscience* 3, no. 12 (2010): 813–16.

Hsu, Chia-Wei, and Isabella Velicogna. "Detection of Sea Level Fingerprints Derived from GRACE Gravity Data." *Geophysical Research Letters* 44, no. 17 (2017): 8953–61.

Hughes, Paul, and Alan D. Wall. "The Admiralty Tidal Predictions of 1833: Their Comparison with Contemporary Observation and with a Modern Synthesis." *Journal of Navigation*, no. 57 (2004): 203–14.

Humboldt, Alexander von. *Asie centrale: Recherches sur les chaines de montagnes et la climatologie comparée*. Vol. 2. Paris: Gide, 1843.

———. *Essai politique sur le royaume de la Nouvelle Espagne*. Vol. 1. Paris: Schoell, 1811.

———. "On the Comparative Level of Lakes and Seas, as the Caspian, Lake Aral, Black Sea, Red Sea, Mediterranean, Lake Tiberias, Dead Sea, Atlantic, Pacific, &c." *Edinburgh New Philosophical Journal* 35 (1843): 323–35.

———. *Relation historique du voyage aux régions équinoxiales du nouveau continent*. Vol. 3. 1825. Reprint, Stuttgart: Brockhaus, 1970.

Hunchuck, Elise. "An Incomplete Atlas of Stones: A Cartography of the Tsunami Stones on the Japanese Shoreline." *Funambulist*, July 7, 2018. https://thefunambulist.net/magazine/cartography-power/incomplete-atlas-stones-cartography-tsunami-stones-japanese-shoreline-elise-misao-hunchuck.

Hunter, J., R. Coleman, and D. Pugh. "The Sea Level at Port Arthur, Tasmania, from 1841 to the Present." *Geophysical Research Letters* 30, no. 7 (2003).

Huq, Saleemul, Syed Iqbal Ali, and A. Atiq Rahman. "Sea-Level Rise and Bangladesh: A Preliminary Analysis." *Journal of Coastal Research*, no. 14 (1995): 44–53.

Ibáñez e Ibáñez de Ibero, Carlos. "Rapport sur l'état des travaux faits pour la détermination du niveau moyen des mers de l'Europe continentale." In *Verhandlungen der vom 13. bis zum 16. September 1880 zu München abgehaltenen sechsten allgemeinen Conferenz der Europäischen Gradmessung*, edited by Adolph Hirsch, Annexe VIII. Berlin: Georg Reimer, 1881.

———. "Rapport sur l'état des travaux faits pour la détermination du niveau moyen des mers de l'Europe continentale." In *Verhandlungen der vom 15. bis zum 24. October 1883 in Rom abgehaltenen siebenten allgemeinen Conferenz der Europäischen Gradmessung*, edited by Adolph Hirsch, Annexe V. Berlin: Georg Reimer, 1884.

———. "Rapport sur l'état des travaux faits pour la détermination du niveau des mers de l'Europe continentale." In *Verhandlungen der vom 3. bis 12. October 1889 in Paris abgehaltenen neunten allgemeinen Conferenz der Internationalen Erdmessung und deren Permanenten Commission*, edited by Adolph Hirsch, Annexe A II. Berlin: Georg Reimer, 1890.

IGN–Institut National de l'Information Géographique et Forestiére. "Les réseaux de nivellement français," July 9, 2012. https://geodesie.ign.fr/index.php?page=reseaux_nivellement_francais.

Imbrie, John, and Katherine Palmer Imbrie. *Ice Ages: Solving the Mystery*. Cambridge, MA: Harvard University Press, 1979.

"Influence de la pression atmosphérique dur le niveau moyen de la mer; par M. Daussy." *Comptes rendus hebdomadaires des seances de l'academie des sciences* 3 (1836): 136–38.

IPCC. "Global Warming of 1.5°." Special Report. Accessed October 14, 2021, https://www.ipcc.ch/sr15/.

Ireton, Sean, and Caroline Schaumann. "Introduction: The Meaning of Mountains: Geology, History, Culture." In *Heights of Reflection: Mountains in the German Imagination from the Middle Ages to the Twenty-First Century*, edited by Sean Ireton and Caroline Schaumann, 1–19. Studies in German Literature, Linguistics, and Culture. Rochester, NY: Camden House, 2012.

Irons, James Campbell, ed. *Autobiographical Sketch of James Croll . . . with Memoir of His Life and Work*. London: Edward Stanford, 1896.

Ismail-Zadeh, Alik, and Jo Ann Joselyn. "IUGG: Beginning, Establishment, and Early Development (1919–1939)." *History of Geo- and Space Sciences* 10, no. 1 (2019): 25–44.

Jacchia, L. G. "The Earth's Gravitational Potential as Derived from Satellites 1957 Beta One and 1958 Beta Two." *SAO Special Report* 19 (1958): 19.

James, Henry. *Abstracts of the Principal Lines of Spirit Levelling in England and Wales*. London: G. E. Eyre and W. Spottiswoode, 1861.

Jelgersma, S. *Holocene Sea Level Changes in the Netherlands*. Maastricht: van Aelst, 1961.

Johnson, Douglas W. *Fixité de la Côte Atlantique de l'Amerique du Nord*. Paris: Librairie Armand Colin, 1912.

———. "Is the Atlantic Coast Sinking?" *Geographical Review* 3, no. 2 (1917): 135–39.

Johnson, Douglas, and Elizabeth Winter. "Sea-Level Surfaces and the Problem of Coastal Subsidence." *Proceedings of the American Philosophical Society* 66 (1927): 465–96.

Johnson, Nathanael. "What the Tiny Tangier Island Teaches Us about Sea Level Rise." *Grist*, November 10, 2021. https://grist.org/article/what-a-tiny-island-in-chesapeake-bay-teaches-us-about-the-costs-of-sea-level-rise/.

Jones, E. Lester. *Use of Mean Sea Level as the Datum for Elevations*. Special Publications 41. Washington DC: Government Printing Office, 1917.

Juan, Jorge, and Antonio de Ulloa. *Voyage historique de l'Amerique meridionale*. Vol. 2. Amsterdam and Leipzig: Arkstee & Merkus, 1752.

Kaula, William M. *Theory of Satellite Geodesy: Applications of Satellites to Geodesy*. Waltham, MA: Blaisdell, 1966.

Kaye, Clifford A. "The Upper Limit of Barnacles as an Index of Sea-Level Change on the New England Coast during the Past 100 Years." *Journal of Geology* 72, no. 5 (1964): 580–600.

Khandker, Hasina. "Mean Sea Level in Bangladesh." *Marine Geodesy* 20 (1994).

Kirrinnis, H. "Friedrich Wilhelm Bessel und seine Bedeutung für die Geographie." *Erdkunde* 5, no. 3 (1951): 247–49.

Knowles, Scott Gabriel. "Slow Disaster in the Anthropocene: A Historian Witnesses Climate Change on the Korean Peninsula." *Daedalus* 149, no. 4 (2020): 192–206.

Koop, Frederick W. *Precise Leveling in New York City*. New York: City of New York, Board of Estimate and Apportionment, 1915.

Kortum, Gerhard. "The Naval Observatory in Tsingtau (Qingdao), 1897–1914: German Background and Influence." In *Ocean Sciences Bridging the Millenia—A Spectrum of Historical Accounts*, edited by S. Morcos, M. Zhu, R. Charlier, M. Gerges, G. Kullenberg, W. Lenz, M. Lu, E. Zou, and G. Wright, 251–67. Paris: UNESCO Publishing, 2004.

Krajic, Kevin. "Wallace Broecker, Prophet of Climate Change." *State of the Planet* (blog), February 19, 2019. https://news.climate.columbia.edu/2019/02/19/wallace-broecker-early-prophet-of-climate-change/.

Krüger, Tobias. *Discovering the Ice Ages: International Reception and Consequences for a Historical Understanding of Climate*. Leiden: Brill, 2013.

Kuenen, P. H. *Marine Geology*. New York: J. Wiley & Sons, 1950.

Kutazov, I. A. "Scientific Problems of Geodesy and Cartography." In *Vestnik Akademii Nauk SSSR*, vol. 42, no. 9 (1972): 37–43. JPRS 57455. Arlington, VA: Joint Publications Research Service, 1972.

Lachapelle, Gérard. "Status of the Redefinition of the Vertical Reference System in Canada." In *Proceedings: Second International Symposium on Problems Related to the Redefinition of North American Geodetic Networks*, 609–15. US Department of Commerce, National Oceanic and Atmospheric Administration, National Ocean Survey, 1978.

Lalande, Joseph Jérôme Le Français de. *Traité du flux et du reflux de la mer*. Paris: Veuve Desaint, 1781.

Lallemand, Charles. "L'unification des altitudes et le niveau des mers en Europe." In *Association française pour l'avancement des sciences: Compte rendu de la 19me session*, 930–40. Paris: M. G. Masson, 1891.

———. "Note sur le principe fondamental de la théorie du nivellement." In *Verhandlungen der vom 27. October bis zum 1. November 1886 in Berlin abgehaltenen achten allgemeinen Conferenz der Internationalen Erdmessung und deren Permanenten Commission*, edited by Adolph Hirsch, Annexe Vb. Berlin: Georg Reimer, 1887.

———. "Note sur les travaux exécutés par le service du Nivellement général de la France en 1889." In *Verhandlungen der vom 3. bis 12. October 1889 in Paris abgehaltenen neunten allgemeinen Conferenz der Internationalen Erdmessung und deren Permanenten Commission*, edited by Adolph Hirsch, Annexe B XVIIIb. Berlin: Georg Reimer, 1890.

———. "Note sur l'unification des altitudes européennes." In *Verhandlungen der vom 15. bis 21. September 1890 zu Freiburg i. B. abgehaltenen Conferenz der Permanenten Commission der Internationalen Erdmessung*, edited by Adolph Hirsch, Annexe C II. Berlin: Georg Reimer, 1891.

———. "Rapport présenté au nom de la Commision du zéro international des altitudes." In *Verhandlungen der vom 12. bis 18. September 1893 in Genf abgehaltenen Conferenz der Permanenten Commission der Internationalen Erdmessung*, edited by Adolph Hirsch, Annexe A III. Berlin: Reimer, 1894.

Lambeck, Kurt, and Richard Coleman. "The Earth's Shape and Gravity Field: A Report of Progress from 1958 to 1982." *Geophysical Journal of the Royal Astronomical Society*, no. 74 (1983): 25–54.

Lambert, Walter D. "The International Geodetic Association (Die Internationale Erdmessung) and Its Predecessors." *Bulletin Géodésique* 17, no. 1 (1950): 299–324.

Lambright, W. Henry. "The Political Construction of Space Satellite Technology." *Science, Technology, & Human Values* 19, no. 1 (1994): 47–69.

Lambton, William. "An Account of the Measurement of an Arc on the Meridian on the Coast of Coromandel, and the Length of a Degree Deduced Therefrom in the Latitude 12d 32f." *Asiatick Researches* 8 (1808): 136–93.

Lane, Alfred C., William Bowie, Roy E. Dickerson, E. A. Eckhardt, C. L. Garner, W. H. Hobbs, W. J. Humphreys, et al. "Round Table Discussion." *Proceedings of the American Philosophical Society* 79, no. 1 (1938): 127–44.

Laplace, Pierre Simon. *Traité de mécanique céleste*. Paris: Imprimerie Crapelet, 1798.

Lehman, Jessica. "From Ships to Robots: The Social Relations of Sensing the World Ocean." *Social Studies of Science* 48, no. 1 (2018): 57–79.

Lenz, Emil. "Beschreibung eines sich selbst registrirenden Fluthmessers, nebst einigen mit diesem Apparate erhaltenen vorläufigen Resultaten." *Annalen der Physik und Chemie* vol. 136, no. 11 (1843): 408–12.

——. "Ueber die Veränderung der Höhe, welche die Oberfläche des Kaspischen Meeres bis zum April des Jahres 1830 erlitten hat." *Annalen der Physik und Chemie*, vol. 102, no. 11 (1832): 353–94.

Le Père, Gratien. "Plan et nivellement des source dites de Moyse." In *Description de l'Égypte, ou, Recueil de observations et des recherches qui ont été faites en Égypte pendant l'éxpédition de l'armée française*, edited by Edme François Jomard. État moderne. Planches. 1:Pl. 10. Paris: Imprimerie Imperiale, 1809.

Le Père, Jacques-Marie. "Mémoire sur la communication de la Mer des Indes à la Mediterranée par la Mer Rouge et l'Isthme de Suez." In *Description de l'Égypte, ou, Recueil de observations et des recherches qui ont été faites en Égypte pendant l'éxpédition de l'armée française*, edited by Edme François Jomard. État moderne. 1:21–31. Paris: Imprimerie Imperiale, 1809.

Lesseps, Ferdinand-Marie de. *Percement de l'Isthme de Suez: rapport et projet de la Commission Internationale*. Paris: H. Plon, 1856.

"Letter from the Baron Alexander von Humboldt to the Earl of Minto, 12 October 1839." In *Report of the Committee of Physics, Including Meteorology, on the Objects of Scientific Inquiry in Those Sciences*, 91–102. London: Richard and John E. Taylor, 1840.

Linke, Robert. "The Influence of German Surveying on the Development of New Guinea." Shaping the Change, XXIII FIG Congress, Munich, 2006. http://www.aspng.org/hs02_04_linke_0976.pdf.

Listing, Johann Benedikt. *Ueber unsere jetzige Kenntniss der Gestalt und Grösse der Erde*. Göttingen: Dieterich, 1872.

Lloyd, John Augustus. "Account of Levellings Carried across the Isthmus of Panama, to Ascertain the Relative Height of the Pacific Ocean at Panama and of the Atlantic at the Mouth of the River Chagres; Accompanied by Geographical and Topographical Notices of the Isthmus." *Philosophical Transactions of the Royal Society of London* 120 (1830): 59–68.

——. "An Account of Operations Carried on for Ascertaining the Difference of Level between the River Thames at London Bridge and the Sea; and Also for Determining the Height above the Level of the Sea, &c. of Intermediate Points Passed Over between Sheerness and London Bridge." *Philosophical Transactions of the Royal Society of London* 121 (1831): 167–97.

Löschner, Hans. "Zur Frage der Vereinheitlichung der Ausgangspunkte der Präzisionsnivellements." *Österreichische Zeitschrift für Vermessungswesen* 4, no. 7–8 (1906): 89–101.

Lubbock, John William. Letter to William Whewell, 1831. ADD.MS.a/208/80. Trinity College Library, Cambridge. https://archives.trin.cam.ac.uk/index.php/john-william-lubbock-to-william-whewell-8.

Lubrich, Oliver. "Fascinating Voids: Alexander von Humboldt and the Myth of Chimborazo." In *Heights of Reflection: Mountains in the German Imagination from the Middle Ages to the Twenty-First Century*, edited by Sean Ireton and Caroline Schaumann, 153–75. Studies in German Literature, Linguistics, and Culture. Rochester, NY: Camden House, 2012.

Lucarelli, Fosco. "The Three Mawangdui Maps: Early Chinese Cartography." *Socks*, March 2, 2014. https://socks-studio.com/2014/03/02/the-three-mawangdui-maps-early-chinese-cartography/.

Lucier, Paul. "The Professional and the Scientist in Nineteenth-Century America." *Isis* 100, no. 4 (2009): 699–732.

Lyell, Charles. "The Bakerian Lecture: On the Proofs of a Gradual Rising of the Land in Certain Parts of Sweden." *Philosophical Transactions of the Royal Society of London* 125 (1835): 1–38.

———. *Principles of Geology*. 1st ed. 3 vols. London: John Murray, 1830.

———. *Principles of Geology*. 9th ed. Boston, MA: Little, Brown, and Company, 1853.

Macfarlane, Robert. "Silt." *Granta*, no. 119 (2012): 41–60.

Maclaren, Charles. "The Glacial Theory of Prof. Agassiz." *American Journal of Science and Arts* 42 (1842): 346–65.

Magnason, Andri Snær. *On Time and Water*. London: Serpents Tail, 2020.

Mahony, Martin. "For an Empire of 'All Types of Climate': Meteorology as an Imperial Science." *Journal of Historical Geography* 51 (2016): 29–39.

Malakhov, B. M. "Marine Geodesy." In *Geodeziya i Kartografiya*, no. 2, pp. 19–25. JPRS 58824. Arlington, VA: Joint Publications Research Service, 1973.

Malcolm, Sir John. *Memoir of Central India*. Vol. 2. London: Kingsbury, Parbury, & Allen, 1823.

"Maldives Sends Climate SOS with Undersea Cabinet," Reuters, October 17, 2009. https://www.reuters.com/article/us-maldives-environment-idUSTRE59G0P120091017.

Marmer, Harry A. "Is the Atlantic Coast Sinking? The Evidence from the Tide." *Geographical Review* 38, no. 4 (1948): 652–57.

———. "The Purpose of Tide Observations." *Scientific Monthly* 35, no. 2 (1932): 162–68.

Martin, Alison E. "Translation, Annotation and Knowledge-Making: Leopold von Buch's *Travels through Norway and Lapland* (1813)." *Comparative Critical Studies* 16, no. 2–3 (2019): 323–41.

Matthäus, Wolfgang. "On the History of Recording Tide Gauges." *Proceedings of the Royal Society of Edinburgh, Section B: Biological Sciences* 73 (1972): 26–34.

Mauelshagen, Franz. "Climate Change, Decline and Societal Collapse in the Writing of History." Declinism Seminar, Engelsberg Ironworks, Sweden, 2015. https://www.researchgate.net/publication/314141866_Climate_Change_Decline_and_Societal_Collapse_in_the_Writing_of_History.

Medvedev, Igor P., Alexander B. Rabinovich, and Evgueni A. Kulikov. "Tides in Three Enclosed Basins: The Baltic, Black, and Caspian Seas." *Frontiers in Marine Science* 3 (2016). https://www.frontiersin.org/article/10.3389/fmars.2016.00046.

Menefee, Samuel Pyeatt. "'Half Seas Over': The Impact of Sea Level Rise on International Law and Policy." *UCLA Journal of Environmental Law and Policy* 9, no. 2 (1991).

Meyer, H. A., K. Möbius, G. Karsten, and V. Hensen. *Jahresbericht der Commission zur Wissenschaftlichen Untersuchung der Deutschen Meere in Kiel für das Jahr 1871*. Berlin: Wiegandt & Hempel, 1873.

Middleton, W. E. Knowles. *The History of the Barometer*. Baltimore: Johns Hopkins, 1964.

Miller, Sarah. "Heaven or High Water: Selling Miami's Last 50 Years." *Popula*, April 2, 2019. https://popula.com/2019/04/02/heaven-or-high-water/.

Ministère de l'Agriculture, du Commerce et des Travaux publics. *Nivellement de la France: Résultats des opérations exécutées pour l'établissement du réseau des lignes de Basebase*. Vol. 1. Bourges: Imprimerie de E. Pigelet, 1864.

Misra, Tanvi. "A Catastrophe for Houston's Most Vulnerable People." *Atlantic*, August 27, 2017. https://www.theatlantic.com/news/archive/2017/08/a-catastrophe-for-houstons-most-vulnerable-people/538155/.

Mitrovica, Jerry X., Natalya Gomez, and Peter U. Clark. "The Sea-Level Fingerprint of West Antarctic Collapse." *Science* 323, no. 5915 (2009): 753.

Montel, Nathalie. "Établir la vérité scientifique au XIXe siècle: La controverse sur la différence de niveau des deux mers (1799–1869)." *Genèses* 32, no. 1 (1998): 86–109.

Moray, Robert "Considerations and Enquiries Concerning Tides, by Sir Robert Moray; Likewise for a Further Search into Dr. Wallis's Newly Publish't Hypothesis." *Philosophical Transactions of the Royal Society of London* 1, nos. 1–22 (1665): 298–301.

Morrell, Jack, and Arnold Thackray. *Gentlemen of Science: Early Year of the British Association for the Advancement of Science*. Oxford: Clarendon Press, 1981.

Moser, Jana. "Untersuchungen zur Kartographiegeschichte von Namibia." Dr.-Ing., TU Dresden, 2007. https://e-docs.geo-leo.de/bitstream/handle/11858/00-1735-0000-0001-3242-2/Moser2007.pdf?sequence=1.

Moule, A. C. "The Bore on the Ch'ien-t'ang River in China." *T'oung Pao* 22, no. 3 (1923): 135–88.

Mudge, William. *An Account of the Operations Carried Out for Accomplishing a Trigonometrical Survey of England and Wales*. 3 vols. London: W. Bulmer and Co., 1801.

Munk, Walter. "Oceanography before, and after, the Advent of Satellites." In *Satellites, Oceanography and Society*, edited by David Halpern. Amsterdam: Elsevier, 2000.

Munk, Walter, and Roger Revelle. "On the Geophysical Interpretation of Irregularities in the Rotation of the Earth." *Geophysical Journal International* 6, no. 6 (1952): 331–47.

———. "Sea Level and the Rotation of the Earth." *American Journal of Science* 250, no. 11 (1952): 829–33.

Nail, Thomas. *Theory of the Earth*. Stanford, CA: Stanford University Press, 2021.

NASA Global Climate Change: Vital Signs of the Planet. "Sea Level," August 2021. https://climate.nasa.gov/vital-signs/sea-level.

National Environmental Satellite, Data, and Information Service. "JASON-3 Mission." Accessed November 3, 2023. https://www.nesdis.noaa.gov/current-satellite-missions/currently-flying/jason-3/jason-3-mission.

National Oceanic and Atmospheric Administration. "Sea Level Rise Viewer v 3.0.0," August 17, 2020. https://coast.noaa.gov/slr/#/layer/slr.

Needham, Joseph. *Science and Civilization in China*. Vol. 3, *Mathematics and the Sciences of the Heavens and the Earth*. Cambridge: Cambridge University Press, 1959.

Nienhuis, Piet H. *Environmental History of the Rhine–Meuse Delta*. Dordrecht: Springer Netherlands, 2008.

Nunn, Patrick D., and Nicholas J. Reid. "Aboriginal Memories of Inundation of the Australian Coast Dating from More than 7000 Years Ago." *Australian Geographer* 47, no. 1 (2016): 11–47.

Observations of the Tides: Communicated to the Royal Society by the Admiralty. London: Richard Taylor, 1833.

Ogle, Vanessa. *The Global Transformation of Time: 1870–1950*. Cambridge, MA: Harvard University Press, 2015.

O'Hara, Kieran D. *A Brief History of Geology*. Cambridge: Cambridge University Press, 2018.

Ohnesorge, Miguel. "How Incoherent Measurement Succeeds: Coordination and Success in the Measurement of the Earth's Polar Flattening." *Studies in History and Philosophy of Science* 88 (2021): 245–62.

———. "The Promises and Pitfalls of Precision: Measurement and Systematic Error in Physical Geodesy, c. 1800–1910." *Annals of Science* 81, nos. 1–2 (2024): 258–84.

Palmer, Henry Robinson, and John William Lubbock. "Description of a Graphical Registrer of Tides and Winds." *Philosophical Transactions of the Royal Society of London* 121 (1831): 209–13.

Pasumot, François. "Lettre aux auteurs du Journal de Physique." *Observations sur la physique, sur l'histoire naturelle et sur les arts* 23 (1783): 193–201.

Peirce, Charles S. Letter to Sarah Mills, November 2, 1877. Correspondencia europea de C. S. Peirce: creatividad y cooperación científica. Universidad de Navarra. https://www.unav.es/gep/Havre02.11.1877En.html.

Penck, Albrecht. "Eustatische Bewegungen des Meeresspiegels während der Eiszeit." *Geographische Zeitschrift* 39, no. 6 (1933): 329–39.

———. *Schwankungen des Meeresspiegels*. Separatabdruck aus dem Jahrbuch der Geographischen Gesellschaft zu München 7. Munich: Theodor Ackermann, 1882.

Permanent Service for Mean Sea Level (PSMSL). "Tide Gauge Data," 2017. http://www.psmsl.org/data/obtaining/.

Personnaz, Charlene, and Amelie Herenstein. "Mont Blanc Shrinks by Over Two Meters in Two Years." *phys.org* (blog), October 5, 2023. https://phys.org/news/2023-10-mont-blanc-meters-years.html.

Pethick, J., and J. Orford. "Rapid Rise in Effective Sea-Level in Southwest Bangladesh: Its Causes and Contemporary Rates." *Global and Planetary Change* 111 (2013): 237–45.

Phillimore, Reginald H. *Historical Records of the Survey of India*. 5 vols. Dehra Dun: Survey of India, 1945–1968.

Playfair, John. *Illustrations of the Huttonian Theory of the Earth*. Edinburgh: William Creech, 1802.

Polezhayev, A. P. "Use of Artificial Earth Satellites for Geodesy." *Izvestiya VUZ, Geodeziya i Aerofotos'yemka (News of Higher Educational Institutions, Geodesy and Aerial Photography)*, Moscow, no. 2 JPRS-38537 (1966): 1–31.

Porter, Theodore M. *Trust in Numbers: The Pursuit of Objectivity in Science and Public Life*. Princeton, NJ: Princeton University Press, 1995.

Post, Lennart von. "A Gothiglacial Transgression of the Sea in South Sweden." *Geografiska Annaler* 15, nos. 2–3 (1933): 225–54.

Pouvreau, Nicolas. "Trois cents ans de mesures marégraphiques en France: Outils, méthodes et tendances des composantes du niveau de la mer au port de Brest." PhD diss., Université de La Rochelle, 2008. https://tel.archives-ouvertes.fr/tel-00353660/document.

"Proceedings of the Geological Society—November 11th, 1868." *Quarterly Journal of the Geological Society of London* 25 (1869): 1–12.

Pugh, D. T. "Improving Sea Level Data." In *Climate and Sea Level Change: Observations, Projections and Implications*, edited by R. A. Warrick, E. M. Barrow, and T. M. L. Wigley, 57–71. Cambridge: Cambridge University Press, 1993.

Quell, Molly. "Island Nations Demand Climate Change Action at International Sea Tribunal." Courthouse News Service, September 11, 2023. https://www.courthousenews.com/island-nations-demand-climate-change-action-at-international-sea-tribunal/.

Rabino, Joseph. "The Statistical Story of the Suez Canal." *Journal of the Royal Statistical Society* 50, no. 3 (1887): 495–546.

Rahmstorf, Stefan. "Rising Hazard of Storm-Surge Flooding." *Proceedings of the National Academy of Sciences* 114, no. 45 (2017): 11806–8.

Ramesh, Aditya, and Vidhya Raveendranathan. "Infrastructure and Public Works in Colonial India: Towards a Conceptual History." *History Compass* 18, no. 6 (2020): e12614.

Ramsay, Wilhelm. "Changes of Sea-Level Resulting from the Increase and Decrease of Glaciations." *Fennia* 52, no. 5 (1930).

———. *On Relations between Crustal Movements and Variations of Sea-Level during the Late Quaternary Time Especially in Fennoscandia*. Bulletin de La Commission Géologique de Finlande 66. Helsinki: Imprimerie de l'État, 1924.

Rappleye, Howard S. "The Sea-Level Datum of 1929." *Transactions of the American Geophysical Union* 19, no. 1 (1938): 51–55.

Ravenstein, E. G. "On Bathy-Hypsographical Maps; With Special Reference to a Combination of the Ordnance and Admiralty Surveys." *Proceedings of the Royal Geographical Society and Monthly Record of Geography* 8, no. 1 (1886): 21–27.

Redmount, Carol A. "The Wadi Tumilat and the 'Canal of the Pharaohs'." *Journal of Near Eastern Studies* 54, no. 2 (1995): 127–35.

Regnauld, Hervé, and Patricia Limido. "Coastal Landscape as Part of a Global Ocean: Two Shifts: Images, Ocean, Global Climate." *Geo: Geography and Environment* 3, no. 2 (2016): e00029.

Reid, Clement. *Submerged Forests*. Cambridge: Cambridge University Press, 1913.

Reidy, Michael S. "Gauging Science and Technology in the Early Victorian Era." In *The Machine in Neptune's Garden: Historical Perspectives on Technology and the Marine Environment*, edited by Helen M. Rozwadowski and David K. Van Keuren, 1–38. Sagamore Beach, MA: Science History Publications, 2004.

———. *Tides of History: Ocean Science and Her Majesty's Navy*. Chicago: University of Chicago Press, 2008.

Reidy, Michael S., and Helen M. Rozwadowski. "The Spaces in Between: Science, Ocean, Empire." *Isis* 105, no. 2 (2014): 338–51.

Revelle, Roger. "Oceanography from Space." *Science* 228, no. 4696 (1985): 133–33.

Rich, Nathaniel. "Losing Earth: The Decade We Almost Stopped Climate Change." *New York Times Magazine*, August 1, 2018. https://www.nytimes.com/interactive/2018/08/01/magazine/climate-change-losing-earth.html.

Robert, Eugène. "Recueil d'observations ou recherches géologiques, tendant à prouver, sinon que la mer a baissé et baisse encore de niveau sur tout le globe, notamment dans l'hémisphère nord, du moin que le phénomène de soulèvement, depuis l'époque où il a donné naissance aux grandes chaînes de montagnes, n'a plus guère continué à se manifester que d'une manière lente et graduelle." *Compte rendu des séances de l'Académie des Sciences* 19 (1844): 265–67.

Ross, James Clark. *A Voyage of Discovery and Research in the Southern and Antarctic Regions, during the Years 1839–43*. London: J. Murray, 1847.

Rossi, Francesco Carlo. "Il comune marino di Venezia." In *L'ingegneria a Venezia nell'ultimo ventennio*, 42–48. Venezia: Naratovich, 1887.

Rossiter, J. R. "Les travaux du Service Permanent du Niveau Moyen de la Mer." *Revue Hydrographique Internationale*, no. 40, no. 1 (1963): 85–90.

Royal Commission on Coast Erosion and Afforestation. *Third (and Final) Report of the Royal Commission Appointed to Inquire into and Report on Certain Questions Affecting Coast Erosion: The Reclamation of Tidal Lands, and Afforestation in the United Kingdom*. London: His Majesty's Stationery Office, 1911.

Rozwadowski, Helen M. *Fathoming the Ocean: The Discovery and Exploration of the Deep Sea*. Cambridge, MA: Belknap, 2005.

Rudwick, Martin J. S. *Bursting the Limits of Time: The Reconstruction of Geohistory in the Age of Revolution*. Chicago: University of Chicago Press, 2005.

———. *Worlds before Adam: The Reconstruction of Geohistory in the Age of Reform*. Chicago: University of Chicago Press, 2008.

Sammler, Katherine G. "Kauri and the Whale: Oceanic Matter and Meaning in New Zealand." In *Blue Legalities: The Life and Laws of the Sea*, edited by Irus Braverman and Elizabeth R. Johnson, 63–84. Durham, NC: Duke University Press, 2020.

———. "The Rising Politics of Sea Level: Demarcating Territory in a Vertically Relative World." *Territory, Politics, Governance* 8, no. 5 (2019): 604–20.

Sánchez, Laura, Jonas Ågren, Jianliang Huang, Yan Ming Wang, Jaakko Mäkinen, Roland Pail, Riccardo Barzaghi, Georgios S. Vergos, Kevin Ahlgren, and Qing Liu. "Strategy for the Realisation of the International Height Reference System (IHRS)." *Journal of Geodesy* 95, no. 3 (2021): 33.

Saussure, Horace-Benedicte de. *Relation abrégée d'un voyage à la cime du Mont-Blanc. En août 1787*. Geneva: Barde, Manget & Compagnie, 1787.

Schaffer, Simon. "Oriental Metrology and the Politics of Antiquity in Nineteenth-Century Survey Sciences." *Science in Context* 30, no. 2 (2017): 173–212.

Scheuchzer, Johann Jakob. *Nova Helvetiae tabula geographica*. 1:230,000. Zurich, 1712. Ryh 8608: 4–7. Universitätsbibliothek Bern. https://biblio.unibe.ch/maps/ryh/ch/rec/r0000914.htm.

Scheuchzer, Johannes Gaspar. "The Barometrical Method of Measuring the Height of Mountains, with Two New Tables Shewing the Height of the Atmosphere at Given Altitudes of Mercury. Extracted Chiefly from the Observations of John James Scheuchzer, M. D. Professor of Mathematics at Zurich, and a Member of the Imperial and Royal Societies of London and Prussia." *Philosophical Transactions of the Royal Society* 35, no. 405 (1728): 537–47.

Schofield, Clive. "Holding Back the Waves? Sea-Level Rise and Maritime Claims." In *Climate Change: International Law and Global Governance*, edited by Oliver C. Ruppel, Christian Roschmann, and Katharina Ruppel-Schlichting, 1:593–614. Baden-Baden: Nomos, 2013.

Schwartz, Robert, Ian Gregory, and Thomas Thévenin. "Spatial History: Railways, Uneven Development, and Population Change in France and Great Britain, 1850–1914." *Journal of Interdisciplinary History* 42, no. 1 (2011): 53–88.

Scott, James C. *Seeing Like a State: How Certain Schemes to Improve the Human Condition Have Failed*. New Haven, CT: Yale University Press, 1998.

Section d'oceanographie physique. *Réunion plénière de la section (Madrid, Octobre 1924)*. Bulletin 5. Venezia: Premiate Officine Grafiche Carlo Ferrari, 1925.

———. *Réunion plénière de la section (Prague, Septembre 1927)*. Bulletin 11. Venezia: Premiate Officine Grafiche Carlo Ferrari, 1928.

———. *Réunion plénière de Paris (25–28 Janvier 1921)*. Bulletin 1. Venezia: Premiate Officine Grafiche Carlo Ferrari, 1921.

———. *Réunion plénière: Rome (4–10 Mai 1922)*. Bulletin 3. Venezia: Premiate Officine Grafiche Carlo Ferrari, 1923.

Seibold, E., and I. Seibold. "Vereisung und Meeresspiegel." *Geologische Rundschau* 85, no. 3 (1996): 403–8.

Shabecoff, Philip. "Haste of Global Warming Trend Opposed." *New York Times*, October 21, 1983. https://www.nytimes.com/1983/10/21/us/haste-of-global-warming-trend-opposed.html.

Shaler, Nathaniel. "Notes on Some of the Phenomena of Elevation and Subsidence of the Continents." *Proceedings of the Boston Society of Natural History* 17 (1875): 288–92.

Shapin, Steven. "The Invisible Technician." *American Scientist* 77, no. 6 (1989): 554–63.

Shapin, Steven, and Simon Schaffer. *Leviathan and the Air-Pump: Hobbes, Boyle, and the Experimental Life*. Princeton, NJ: Princeton University Press, 1985.

Shepard, F. P., and H. E. Suess. "Rate of Postglacial Rise of Sea Level." *Science* 123, no. 3207 (1956): 1082–83.

Shum, C. K., J. C. Ries, and B. D. Tapley. "The Accuracy and Applications of Satellite Altimetry." *Geophysical Journal International* 121, no. 2 (1995): 321–36.

Sieger, Robert. "Seenschwankungen und Strandverschiebungen in Skandinavien." *Zeitschrift der Gesellschaft für Erdkunde zu Berlin* 28, no. 1 (1893): 1–106.

Siegert, Bernhard. "Cacography or Communication? Cultural Techniques in German Media Studies." *Grey Room* 29 (2007): 26–47.

Simpson, George Gaylord. "Uniformitarianism: An Inquiry into Principle, Theory, and Method in Geohistory and Biohistory." In *Essays in Evolution and Genetics in Honor of Theodosius Dobzhansky: a Supplement to Evolutionary Biology*, edited by Max K. Hecht and William C. Steere, 43–96. Boston: Springer US, 1970.

Skelton, R. A. "Cartography." In *A History of Technology*, edited by Charles Singer, E. J. Holmyard, A. R. Hall, and Trevor I. Williams, 4:596–628. Oxford: Clarendon Press, 1958.

SMHI. "Tides," April 23, 2014. https://www.smhi.se/en/theme/tides-1.11272.

Sobel, Dava. *Longitude: The True Story of a Lone Genius Who Solved the Greatest Scientific Problem of His Time*. New York: Walker, 1995.

Sörlin, Sverker. "The Anxieties of a Science Diplomat: Field Coproduction of Climate Knowledge and the Rise and Fall of Hans Ahlmann's 'Polar Warming.'" *Osiris* 26, no. 1 (2011): 66–88.

———. "The Global Warming That Did Not Happen: Historicizing Glaciology and Climate Change." In *Nature's End: History and the Environment*, edited by Sverker Sörlin and Paul Warde, 93–114. London: Palgrave Macmillan UK, 2009.

Sörlin, Sverker, and Nina Wormbs. "Environing Technologies: A Theory of Making Environment." *History and Technology* 34, no. 2 (2018): 101–25.

Spata, Manfred. "Historische Pegel und Bezugshöhen in Europa." *Deutsches Schiffahrtsarchiv*, no. 21 (1998): 379–92.

Spencer, Tom. "Glacial Control Hypothesis." In *Encyclopedia of Modern Coral Reefs: Structure, Form and Process*, edited by David Hopley, 486–91. Dordrecht: Springer Netherlands, 2011.

"Standards and Procedures for Referencing Project Elevation Grades to Nationwide Vertical Datums." Manual No. EM 1110-2-6056. Washington, DC: US Army Corps of Engineers, 2010. https://www.publications.usace.army.mil/portals/76/publications/engineermanuals/em_1110-2-6056.pdf.

Steffen, Will, Wendy Broadgate, Lisa Deutsch, Owen Gaffney, and Cornelia Ludwig. "The Trajectory of the Anthropocene: The Great Acceleration." *Anthropocene Review* 2, no. 1 (2015): 81–98.

Strauss, Benjamin H., Philip M. Orton, Klaus Bittermann, Maya K. Buchanan, Daniel M. Gilford, Robert E. Kopp, Scott Kulp, Chris Massey, Hans de Moel, and Sergey Vinogradov. "Economic Damages from Hurricane Sandy Attributable to Sea Level Rise Caused by Anthropogenic Climate Change." *Nature Communications* 12, no. 1 (2021): 2720.

Suess, Eduard. *Das Antlitz der Erde*. Vol. 2. Wien: F. Tempsky, 1888.

———. *Erinnerungen*. Leipzig: Verlag von S. Hirzel, 1916.

———. "Untersuchungen über den Charakter der österreichischen Tertiärablagerung—I. Über die Gliederung der tertiären Bildungen zwischen dem Mannhart, der Donau und dem äusseren Saume des Hochgebirges." *Sitzungsberichte der Akademie der Wissenschaften. Mathematisch-naturwissenschaftliche Klasse* 54 (1866): 87–149.

Suess, Franz E. "Zur Deutung der Vertikalbewegungen der Festländer und Meere." *Geologische Rundschau* 11, nos. 1–4 (1920): 144–68.

———. "Zur Deutung der Vertikalbewegungen der Festländer und Meere." *Geologische Rundschau* 11, no. 7 (1921): 361–406.

Sullivan, Walter. "So Far, Greenhouse Effect Heats Only Debate." *New York Times*, October 23, 1983, sec. Week in Review. https://www.nytimes.com/1983/10/23/weekinreview/so-far-greenhouse-effect-heats-only-debate.html.

Sweet, W. V., R. Horton, R. E. Kopp, A. N. LeGrande, and A. Romanou. "Sea Level Rise." Chap. 12 in *Climate Science Special Report: Fourth National Climate Assessment*, edited by David J. Dokken, David W. Fahey, Kathy A. Hibbard, Thomas K. Maycock, Brooke C. Stewart, and Donald J. Wuebbles, 1:333–62. Washington, DC: US Global Change Research Program, 2017.

Tamisiea, Mark E., Chris W. Hughes, Simon D. P. Williams, and Richard M. Bingley. "Sea Level: Measuring the Bounding Surfaces of the Ocean." *Philosophical Transactions of the Royal Society A: Mathematical, Physical and Engineering Sciences* 372, no. 2025 (2014): 20130336.

Teller, Edward. "Energy Patterns of the Future." In *Energy and Man: A Symposium*, by Allan Nevins, Robert G. Dunlop, Edward Teller, Edward S. Mason, and Herbert Hoover Jr., 53–72. New York: Appleton-Century-Crofts, 1960.

Theberge, Albert E. "150 Years of Tides on the Western Coast: The Longest Series of Tidal Observations in the Americas." National Oceanic and Atmospheric Administration, 2005. https://tidesandcurrents.noaa.gov/publications/150_years_of_tides.pdf.

Thompson, D'Arcy W. "On Mean Sea Level and Its Fluctuations." *Fishery Board of Scotland. Scientific Investigations* 1914, no. 4 (1915): 3–45.

Thomson, William. "Polar Ice-Caps and Their Influence in Changing Sea Levels." *Transactions of the Geological Society of Glasgow* 8, no. 2 (1888): 322–40.

Thorarinsson, Sigurdur. "Present Glacier Shrinkage, and Eustatic Changes of Sea-Level." *Geografiska Annaler* 22 (1940): 131–59.

"Tide Gauge at Sheerness, The." *Nautical Magazine* 1, no. 8 (1832): 401–4.

Tiny Tim. *The Other Side*. Reprise Records, 1968. https://genius.com/Tiny-tim-the-other-side-lyrics.

Tingle, Alex. "Flood Maps." Accessed October 14, 2021. http://flood.firetree.net/?ll=52.6813,5.5283&zoom=7&m=1.

———. "More about Flood Maps." *Firetree.net* (blog), May 18, 2006. http://blog.firetree.net/2006/05/18/more-about-flood-maps/.

Tirkot, Ursula. "Bibliothek des GeoForschungszentrum (GFZ)." In *Handbuch der historischen Buchbestände in Deutschland, Österreich und Europa*, edited by Bernhard Fabian. Hildesheim: Olms Neue Medien, 2003. https://fabian.sub.uni-goettingen.de/fabian?GeoForschungszentrum_(GFZ).

Tollefson, Jeff. "Satellite Snafu Masked True Sea-Level Rise for Decades." *Nature News* 547, no. 7663 (2017): 265.

Torge, Wolfgang. "From a Regional Project to an International Organization: The 'Baeyer-Helmert-Era' of the International Association of Geodesy 1862–1916." In *IAG 150 Years: Proceedings of the 2013 IAG Scientific Assembly, Potsdam,Germany, 1–6 September, 2013*, edited by Chris Rizos and Pascal Willis, 3–18. Cham: Springer, 2016.

———. *Geschichte der Geodäsie in Deutschland*. Berlin: De Gruyter, 2007.

Trautschold, Hermann. *Sur l'invariabilité du niveau des mers*. Moscow: Khatov, 1879.

———. *Ueber säkulare Hebungen und Senkungen der Erdoberfläche*. Dorpat: C. Mattiesen, 1869.

Trewick, Steven. "Plate Tectonics in Biogeography." In *The International Encyclopedia of Geography*, edited by Douglas Richardson, Noel Castree, Michael F. Goodchild, Audrey Kobayashi, Weidong Liu, and Richard A. Marston, 1–9. London: Wiley & Sons, 2017.

Turcotte, Donald L. *William M. Kaula 1926–2000*. Vol. 81. Biographical Memoirs. Washington, DC: National Academy Press, 2002. http://www.nasonline.org/publications/biographical-memoirs/memoir-pdfs/kaula-william-m.pdf.

"Tuvalu Minister Stands in Sea to Film Cop26 Speech to Show Climate Change," Reuters, November 9, 2021. https://www.reuters.com/business/cop/tuvalu-minister-stands-sea-film-cop26-speech-show-climate-change-2021-11-08/.

Tylor, Alfred. "On Changes of the Sea-Level Effected by Existing Physical Causes during Stated Periods of Time." *Philosophical Magazine* 5, no. 32 (1853): 258–81.

———. "On the Formation of Deltas, and on the Evidence and Cause of Great Changes in the Sea-Level during the Glacial Period." *Geological Magazine* 9, no. 99 (1872): 392–99, 485–501.

Ureta, Sebastián, Thomas Lekan, and Wilko Graf von Hardenberg. "Baselining Nature: An Introduction." *Environment and Planning E: Nature and Space* 3, no. 1 (2020): 3–19.

"Utdrag af Kongl. Vetenskaps Academiens Dag-Bok, samt inkomne Bref och Handlingar." *Kongl. Vetenskaps Academien Handlingar* 18 (1757): 74–77.

Vaníček, Petr. "On the Global Vertical Datum and Its Role in Maritime Boundary Demarcation." *Proceedings of International Symposium on Marine Positioning, INSMAP 94*. Montgomery Village, MD: Marine Technology Society, 1994, 243–50.

Vetter, Jeremy. "Lay Observers, Telegraph Lines, and Kansas Weather: The Field Network as a Mode of Knowledge Production." *Science in Context* 24, no. 2 (2011): 259–80.

Völter, Ulrich. *Geschichte und Bedeutung der Internationalen Erdmessung*. Vol. 63. Deutsche Geodätische Kommission bei der Bayerischen Akademie der Wissenschaften, C. Munich: Verlag der Bayerischen Akademie der Wissenschaften, 1963.

Walker, William. "Observations on the Tides Deduced from Tidal Observations Made in Plymouth Dockyard from August 1831 to September 1833. Condensed from a Lecture Delivered at the Plymouth Institution." *Mechanic's Magazine, Museum, Register, Journal and Gazette* 45 (1846): 245–51, 267–73.

Wanless, Harold R., and Peter Harlem. "A Statement on the Evidence for and Implications of a Recent Rise in Sea Level." Miami: Rosenstiel School of Marine and Atmospheric Science, University of Miami, April 23, 1981.

Warde, Paul, Libby Robin, and Sverker Sörlin. *The Environment: A History of the Idea*. Baltimore: Johns Hopkins University Press, 2019.

Warrick, R., and J. Oerlemans. "Sea Level Rise." In *Climate Change: The IPCC Scientific Assessment*, 257–81. Cambridge: Cambridge University Press, 1990.

Watson, Christopher S., Neil J. White, John A. Church, Matt A. King, Reed J. Burgette, and Benoit Legresy. "Unabated Global Mean Sea-Level Rise over the Satellite Altimeter Era." *Nature Climate Change* 5, no. 6 (2015): 565–68.

Weele, Pieter Izaak van der. *De geschiedenis van het N.A.P.* Delft: Rijkscommissie voor Geodesie, 1971.

"What Happens to a Bridge When One Side Uses Mediterranean Sea Level and Another the North Sea?" *Science 2.0*. October 2, 2013. https://www.science20.com/news_articles/what_happens_bridge_when_one_side_uses_mediterranean_sea_level_and_another_north_sea-121600.

Wheeler, Robert R. "Sanctity of Sea Level." *Geological Society of America Bulletin* 65, no. 12 (1954): 1325.

Whewell, William. "Account of a Level Line, Measured from the Bristol Channel to the English Channel, during the Year 1837–38, by Mr. Bunt." In *Report on the Eighth Meeting of the British Association for the Advancement of Science; Held at Newcastle in August 1838*, 1–11. London: John Murray, 1839.

———. "Researches on the Tides, Seventh Series: On the Diurnal Inequality of the Height of the Tide Especially at Plymouth and at Singapore and on the Mean Level of the Sea." *Philosophical Transactions of the Royal Society* 127 (1837): 75–85.

———. "Researches on the Tides, Tenth Series: On the Laws of Low Water at the Port of Plymouth, and on the Permanency of Mean Water." *Philosophical Transactions of the Royal Society of London* 129 (1839): 151–61.

Whittlesey, Charles. "Depression of the Ocean during the Ice Period." *Proceedings of the American Association for the Advancement of Science* 16 (1868): 92–97.

Wickberg, Adam, and Johan Gärdebo. "Where Humans and the Planetary Conflate—An Introduction to Environing Media." *Humanities* 9, no. 3 (2020): 65.

"Wie hoch liegt ein Ort? Die Verbindung zweier Uhren sagt's mir." Max Planck Institute of Quantum Optics, September 17, 2013. https://www.mpq.mpg.de/4860876/13_09_17.

Wilson, Arnold T. *The Suez Canal*. Oxford: Oxford University Press, 1939.

Wilson, W. S., E. J. Lindstrom, and J. R. Apel. "Satellite Oceanography, History, and Introductory Concepts." In *Encyclopedia of Ocean Sciences*, edited by John H. Steele, 2517–30. Oxford: Academic Press, 2009.

Wise, M. Norton. "Precision: Agent of Unity and Product of Agreement: Part I, Traveling." In *The Values of Precision*, edited by M. Norton Wise, 92–100. Princeton, NJ: Princeton University Press, 1997.

Withers, Charles W. J. *Zero Degrees: Geographies of the Prime Meridian*. Cambridge, MA: Harvard University Press, 2017.

Witting, Rolf. "Mean Sea Level and Its Changes: Unification and Development of the World Net of Mareographs." In *Procès-Verbaux no 1: Cinquième Assemblée Générale Réunie à Lisbonne Septembre 1933*, 41–44. Helsinki: Frenckelska Tryckeri, 1934.

Wolf, Detlef. "The Changing Role of the Lithosphere in Models of Glacial Isostasy: A Historical Review." *Global and Planetary Change* 8, no. 3 (1993): 95–106.

Wood, Gillen D'Arcy. "Climate Delusion: Hurricane Sandy, Sea Level Rise, and 1840s Catastrophism." *Humanities* 8, no. 131 (2019).

Wood, Searles V. "Glacial Submergence." *The Reader* 6, no. 147 (1865): 465–66.

Woodworth, Philip L. "Differences between Mean Tide Level and Mean Sea Level." *Journal of Geodesy* 91, no. 1 (2017): 69–90.

———. "Sea Level Change in Great Britain Between 1859 and the Present." *Geophysical Journal International* 213, no. 1 (2017): 222–36.

———. "A Study of Changes in High Water Levels and Tides at Liverpool during the Last Two Hundred and Thirty Years with Some Historical Background." Birkenhead: Proudman Oceanographic Laboratory, 1999.

———. "Tidal Measurement." In *The History of Cartography*, vol. 6, *Cartography in the Twentieth Century: 1525–28*, edited by Mark Monmonier. Chicago: University of Chicago Press, 2015.

Wöppelmann, Guy, Nicolas Pouvreau, Alain Coulomb, Bernard Simon, and Philip L. Woodworth. "Tide Gauge Datum Continuity at Brest since 1711: France's Longest Sea-Level Record." *Geophysical Research Letters* 35, no. 22 (2008).

Wrede, Ernst Friedrich. *Geognostische Untersuchungen über die Südbaltischen Länder, besonders über das untere Odergebiet*. Berlin: Schüppel, 1804.

Wu, Fumei, Anmin Zeng, and Feng Ming. "Analyzing the Long-Term Changes in China's National Height Datum." *Advances in Space Research* 66 (2020): 1342–50.

Wulf, Andrea. *The Invention of Nature: Alexander von Humboldt's New World*. New York: Knopf, 2015.

Yamarone, Charles A, Sheldon Resell, and David L. Farless. "TOPEX/Poseidon Mapping the Ocean Surface." *Space Congress Proceedings* 4 (1986): 10–22. https://commons.erau.edu/space-congress-proceedings/proceedings-1986-23rd/session-8/4.

Yee, Cordell D. K. "Reinterpreting Traditional Chinese Geographical Maps." In *The History of Cartography*, vol. 2, book 2, edited by J. B Harley and David Woodward, 35–70. Chicago: University of Chicago Press, 1994.

———. "Taking the World's Measure: Chinese Maps between Observation and Text." In *The History of Cartography*, vol. 2, book 2, edited by J. B Harley and David Woodward, 96–127. Chicago: University of Chicago Press, 1994.

Zwingle, Erla. "Watermarks: The Sign of 'C.'" *Venice: I Am Not Making This Up* (blog), August 13, 2009. http://iamnotmakingthisup.net/1988/watermarks-the-sign-of-c/.

INDEX

Page numbers in *italics* refer to illustrations